*Questions and Answers
on
Electronics*

Questions and Answers

on Electronics

Clement Brown

LONDON
NEWNES-BUTTERWORTHS

THE BUTTERWORTH GROUP

ENGLAND
Butterworth & Co (Publishers) Ltd
London: 88 Kingsway, WC2B 6AB

AUSTRALIA
Butterworths Pty Ltd
Sydney: 586 Pacific Highway, NSW 2067
Melbourne: 343 Little Collins Street, 3000
Brisbane: 240 Queen Street, 4000

CANADA
Butterworth & Co (Canada) Ltd
Scarborough: 2265 Midland Avenue, Ontario M1P 4S1

NEW ZEALAND
Butterworths of New Zealand Ltd
Wellington: 26–28 Waring Taylor Street, 1

SOUTH AFRICA
Butterworth & Co (South Africa) (Pty) Ltd
Durban: 152–154 Gale Street

First published in 1966 by George Newnes Ltd
Second impression 1967
Third impression published in 1970 by Newnes-Butterworths,
an imprint of the Butterworth Group
Fourth impression 1972
Fifth impression 1975

ISBN 0 408 00041 4

Made and printed in Great Britain
by Hazell Watson & Viney Ltd,
Aylesbury, Bucks

PREFACE

This condensed account of a wide-ranging technology is intended to give the interested layman an insight into the underlying principles and numerous applications of electronics. Although only a simple explanation of basic electronic phenomena, circuits and components is possible in a book of this size, nevertheless the student will gain from these pages an acquaintance with the "language" of electronics and an understanding of most of the basic operations in this fascinating subject before going on—should he decide to do so—to a more detailed study.

As for the applications of electronic techniques, very many major branches now claim attention, so great is the influence of electronics on activities and industries of all kinds. A number of contrasting topics have been selected and described in the later chapters of this book, to illustrate the practical uses of electronics. On the one hand there is the marriage of electronics to electrical engineering in industrial plant, called control engineering; on the other there are topics like computers and radar, which to the layman are more obviously electronic in character. Radio and television, the most clearly electronic of all, are discussed only briefly as other books in this series provide a more detailed treatment of these subjects.

C.B.

CONTENTS

1

INTRODUCTION

What is meant by electronics ?

Electronics is a technology based on the behaviour, properties and control of electrons. Therefore, although electronics is properly regarded as only a section of electrical technology, electronic techniques are applied in very many fields, including communications, defence, industry and entertainment. Moreover it becomes increasingly difficult to draw clear dividing lines between electronics and other, possibly "heavier" branches of electrical technology: electronic equipment is commonly used in association with industrial electrical systems and power generating plant.

Such a state of affairs reflects the versatility of electronics. However, it must be remembered that the present-day complexity stemmed from something far more limited—the invention of the valve, or electronic tube. In 1883 Edison discovered that under certain conditions electricity would flow in a vacuum. He had experimented with an evacuated electric lamp in which he placed a metal plate. When the plate was made electrically positive with respect to the lamp's filament, a current flowed through the vacuum between filament and plate.

The valve, in its many modern forms, continues to occupy an important place; but it no longer dominates the scene, for it is now joined by other devices, which provide the bases for new electronic techniques. The most obvious and important example is the transistor: it can perform functions similar to those of valves, yet it is unlike the valve physically.

Before the technology of electronics is considered, it is

7

useful to refer briefly to its scientific basis—the nature of the electron and the general properties of conductors, insulators and semiconductors.

What is an electron ?

The electron, found throughout the universe, is the smallest of the fundamental particles. Electrons are fundamental constituents of all matter and form part of all atoms. It is necessary to consider the structure of the atom in order to appreciate the nature of the electron.

The atom consists of a nucleus, about which negatively charged electrons rotate in orbit. The nucleus is made up of positively charged protons and neutral neutrons. Each of these latter particles has a mass nearly the same as that of the hydrogen atom, while the mass of the electron (0·9107 × 10^{-27} gm.) is only $\frac{1}{1840}$ that of the hydrogen atom.

Thus the atom can be regarded as consisting of a nucleus of positive and neutral particles inside a cloud of negatively charged particles—the electrons. (New kinds of particle have been discovered, but in dealing with electronic technology the layman and student need consider only the electrons and the nucleus.) The nucleus accounts for most of the mass of the atom, which is of very open structure.

Of over a hundred elements, hydrogen has the simplest atomic structure: it consists of a single electron moving in orbit around a single proton, as shown in Fig. 1. This will help to convey the idea of an atom, which consists largely of space.

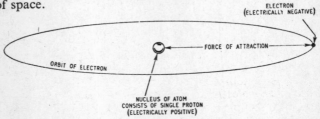

Fig. 1. The hydrogen atom consists largely of space.

As already stated, the electron has a negative charge, and therefore it is attracted to any body which has a positive charge. Ordinarily the positive charge of the atom's other particles balances the electron's charge, so that the atom is electrically neutral in its normal state. However, if an atom loses an electron (through collision between atoms or for other reasons) it has an excess of positive electricity and becomes a *positive ion*; or if it gains an electron it becomes a *negative ion*. In electronics we are concerned with the movement of electrons and ions.

What, then, is an electric current ?

A current is the organised movement of electrons, and the number of electrons moving past a certain point in a given time is the rate of flow of current. In practice we measure current (symbol I) in amperes and (in electronics, particularly) fractions of an ampere. A milliampere is one thousandth of an ampere (abbreviation mA) and a micro-ampere is one millionth of an ampere (μA). Electron current flows round a circuit from the negative to the positive terminal of the source of electrical energy (such as a battery).

What is the difference between conductors and insulators ?

Materials that readily pass electrical currents are known as conductors, and those that resist the flow of current to a marked degree are called insulators. There are also some materials which fall between the extreme cases of good insulators and good conductors: in such instances an electric current flows only with some difficulty. Table 1 compares different materials.

Electrons in an atom have different energies, according to the force needed to remove each electron from the atom. In a solid, consisting of a great many atoms, interplay between adjacent atoms spreads the electron energy into bands. In these bands there are "free electrons"—electrons loosely bound in the atomic structure and therefore free to drift away when suitably influenced.

However, the energy bands are separated by zones which

Table 1. Conductivity of various materials

Some good conductors	Materials which allow some current to flow	Some good insulators
Copper	Carbon	Air
Gold	Copper oxide	Bakelite
Lead	*Germanium	Ceramic
Platinum	Iron	Glass
Silver	Nickel-chrome	Mica
Tin	*Silicon	Dry paper
Salt water		Pure water
		Polystyrene
		Rubber
		Wood

* semiconductors (q.v.)

have no free electrons. The band at the top of this layer arrangement is the conduction band, having the greatest energy. Free electrons in this band move readily when affected by an electric field (e.g. if a voltage is applied), and hence a current can flow through the material.

A good conductor has many free electrons in the conduction band and therefore a large current can flow. In a good insulator, on the other hand, this band has no free electrons and therefore no current can flow. This category embraces both certain natural materials and a large number of more elaborate man-made materials, notably plastics.

What is a semiconductor ?

A number of materials, intermediate in behaviour between conductors and insulators, come under the heading of semiconductors. The discovery and development of such materials have made possible great advances in electronics, especially the transistor and related devices.

As already indicated, some materials (such as most metals) conduct readily because of the presence of free electrons; and insulators prevent current flow because there is a lack of free electrons. The availability of free electrons depends on the material's atomic structure, the relative positions of atoms, and the temperature.

A small number of "foreign" atoms can influence the electrical properties of certain materials. These are the semiconductors, the most important of which at present are germanium and silicon. In general, semiconductors conduct more readily with increase in temperature; the converse is true of good conductors, such as copper. Semiconductors have energy levels below the conduction band mentioned earlier. Their action depends on the movement of *holes*—spaces left by the loss of electrons. A hole is regarded as being opposite in sign to an electron, and, as its name implies, it can accept an electron.

What causes a current to flow ?

Electrons flow in a conductor due to the application of an electromotive force (e.m.f.), which may be provided by a dynamo, alternator, battery or other source. Whatever the source, its influence causes the migration of electrons. The current flow in a conductor is depicted in simple fashion in Fig. 2.

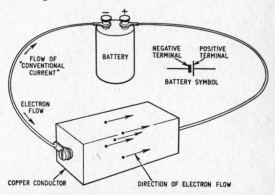

Fig. 2. Current flow in a conductor. The idea of "conventional current" is explained in Section 3.

11

The amount of e.m.f. present between the negative and positive terminals of the source is called the potential difference (p.d.). P.d. and e.m.f. are expressed in volts (symbol E or V).

How are e.m.f. and current related ?

The current in a conductor (that is, the electron flow) depends partly on the size of the e.m.f. and partly on the resistance to flow introduced by the conductor. If the resistance does not change, doubling the e.m.f. will double the current.

In practice all conductors offer some resistance: it may be very small, as in an ordinary copper cable, or it may deliberately be made large, for some specific purpose. Resistance (R) is measured in ohms (Ω), but the higher practical values that are normally encountered in electronics are in thousands of ohms (kΩ) and millions of ohms (megohms, MΩ).

The mathematical relationship between current, potential difference and resistance in a circuit is given by Ohm's Law, thus: $R = V/I$. R is the resistance in ohms, V is the p.d. in volts and I is the current in amperes. The formula can obviously be written in two other ways: $I = V/R$ and $V = RI$.

The resistance of a circuit will have a certain power consumption: energy will be dissipated as heat due to electron collisions in the resistive element. In an electric fire, dissipation of a specified amount of energy will be the aim; whereas in an electric light bulb both light and heat are radiated.

Power in a resistive circuit is expressed as W (watts) $= VI$. (Alternatively $W = I^2R$.) In electronics small powers of thousandths of a watt (milliwatts, mW) are often encountered.

What is the effect of combining resistances ?

If resistors are connected together in series the resistance is increased. In Fig. 3 (a), if both resistors have the same

12

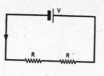
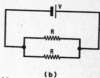

Fig. 3. Resistors connected in series and parallel. With series connection (a) total resistance is 2 × R. With parallel connection (b) the total resistance is R/2.

value R, the resulting resistance will be $2R$. For a given voltage, the current flowing through the combination of two resistors will be half that through one such resistor, R.

If the resistors are connected in parallel, as in Fig. 3 (b), the effective resistance is reduced. In this case, if both resistors have the same value R, the resulting resistance will be $R/2$. For a given voltage the total current flowing through the circuit will be twice that through one resistor of value R.

The question of potential difference across resistors can be introduced at this point. It is obvious that the p.d. across the resistors in Fig. 3 (b) is V. In Fig. 3 (a), however, some of the p.d. "appears" across each resistor. If the resistors are equal, the voltage across each is $V/2$. In practice one resistor might have a large value and the other a small value: a big p.d. would then appear across the large-value resistor, a proportionately smaller voltage appearing across the other.

What is internal resistance ?

So far only the resistance representing a "load" or working circuit element has been considered. However, all sources of e.m.f.—dynamos, oscillators, rectifiers, batteries —have an internal resistance, or source resistance as it is often called.

It is obvious that the current flowing in the external circuit (of Fig. 3, say) also flows through the source of e.m.f.—the battery in this instance. Therefore a little of the e.m.f. is lost as a potential drop across the internal resistance. The greater the flow of current, the greater the lost e.m.f.

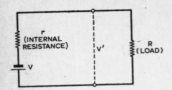

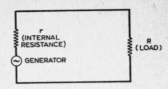

Fig. 4. The internal resistance of the battery is represented by the series resistor r. V' is the actual voltage appearing at the battery output terminals, some of voltage V being lost across r.

Fig. 5. The matched condition is obtained when the load resistance is equal to the internal resistance of the generator.

In Fig. 4 the internal resistance is represented by *r* and the battery has voltage *V*. The load resistance *R* is connected across the output terminals. For a given current, the bigger *r* is the smaller is *V'*, due to loss of volts across *r*.

In practical equipment the internal resistance may be high, and in many instances this will be a disadvantage. Very often, however, a low internal resistance is an important design requirement. For example, a power supply for electronic apparatus will have a very small source resistance so that the output voltage can remain constant over a wide range of currents. Again, the generating plant feeding public and industrial electricity supplies must have a very low internal resistance so that heavy currents can be drawn.

What is the meaning of matching ?

The source of supply and the load are said to be matched when the optimum power is being transferred from one to the other. Of course, some of the power will be dissipated in the internal resistance, as has been noted. We obtain the matched condition when the load resistance is equal to the internal resistance, as indicated in Fig. 5. In electronics there are such requirements as matching an amplifier to other devices or an oscillator to its load.

2

WAVEFORMS, PULSES AND SIGNALS

What is the difference between d.c. and a.c. ?

So far only direct current (d.c.) has been considered: it is so called because it flows in one direction in a circuit. But alternating current (a.c.) is widely used in electronics and indeed in electrical technology generally. Here, the current in the circuit is positive in one direction for a very

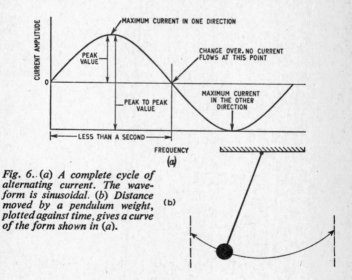

Fig. 6. (a) A complete cycle of alternating current. The waveform is sinusoidal. (b) Distance moved by a pendulum weight, plotted against time, gives a curve of the form shown in (a).

short period; then it falls to zero; and then it increases again in the opposite direction.

A complete change of direction and return to zero is called a cycle. (In U.K. mains supplies these changes occur at the rate of fifty a second.) As indicated in Fig. 6 (a), a complete cycle is a sine wave. There are many opportunities to observe this simple cyclic motion. For instance, the motion of the pendulum in Fig. 6 (b) can easily be converted to the form of Fig. 6 (a): it is necessary only to plot against time the distance moved by the pendulum weight.

What is the advantage of using a.c. ?

The most obvious and important advantage of sinusoidal (sine-wave) current over direct current concerns the use of transformers. A transformer enables voltages and currents to be changed in value and relationship.

The most familiar example of this is of course found in the distribution of electric power for national use. Distribution over long distances is at high voltage and relatively low current, so as to minimise the heating effect (a loss) of very large currents flowing in conductors of practical size. Near the consumers, domestic or industrial, the voltage is reduced to a safe, useable value by transformers, and then large currents can be drawn from the supply.

Apart from power distribution, alternating currents are used in electronics generally—in radio and television, radar, industrial equipment and so on—and without their use many modern developments could not have been achieved. A notable feature is the wide variety of cycle times of these currents. In radio, for example, the cyclic changes of the energy sent out by a transmitting aerial may be at the rate of thousands or millions a second.

What is meant by amplitude ?

The amplitude of an alternating quantity is the maximum departure from the equilibrium value. In Fig. 6 the peak amplitude (current or e.m.f.) and the peak-to-peak amplitude are shown. The root mean square (r.m.s.) value of a sine wave is commonly used: this value is 0·707 times the peak value and is equal to the amplitude of a direct current which, flowing for the same time in the same circuit with the same resistance, would generate the same amount of heat. The amplitude of a wave determines the amount of energy it conveys.

What is frequency ?

Cycle times of alternating currents have already been referred to. The term frequency is used to describe the number of complete cycles occurring in a certain time (normally one second). One cycle a second is called 1 hertz (1Hz). Thus the a.c. mains supply in the U.K. has a frequency of fifty cycles a second, or 50 Hz.

Table 2. Frequency bands

Description	Frequency	Wavelength	Application
Audio frequency (A.F.)	10–30,000 Hz	—	Sound reproduction
Radio frequency: Low frequency (L.F.)	30 kHz to 300 kHz	10,000–1,000 metres	Communications
Medium frequency (M.F.)	300 kHz to 3 MHz	1,000–100 metres	Communications, telemetry
High frequency (H.F.)	3 MHz to 30 MHz	100–10 metres	Communications
Very high frequency (V.H.F.)	30 MHz to 300 MHz	10–1 metres	Communications, television
Ultra high frequency (U.H.F.)	300 MHz to 3,000 MHz	1–0·1 metres	Communications, television
Super high frequency (S.H.F.)	3,000 MHz to 30,000 MHz	100–10 mm	Radar, defence

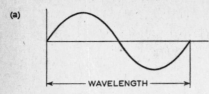

(a)

WAVELENGTH

Fig. 7. Wavelength is the length of a complete cycle, as shown in (a). Equally it is the distance between two similarly placed points on successive cycles, for example A and B in (b).

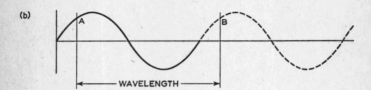

(b)

A

B

WAVELENGTH

In electronics we are concerned with an extremely wide range of frequencies—from a few cycles a second, encountered in sound reproduction and medical electronics, to the many megahertz (millions of cycles a second) of radio communications. This latter term is normally written MHz. The term radio frequency (r.f.) may be applied to energy alternating as low as 10,000 cycles a second (10 kHz).

Frequencies used for communications purposes are divided into bands of frequencies as shown in Table 2.

How is wavelength related to frequency ?

As its name suggests, wavelength is the length of the complete alternating cycle, as shown in Fig. 7 (a). Again, it is the distance between two exactly similarly placed points on successive cycles, such as points A and B in Fig. 7 (b).

This distance is related to frequency by the velocity of the energy concerned. Frequency is numerically equal to velocity divided by wavelength. Therefore wavelength becomes smaller as frequency is raised, and *vice versa*. A sound wave of 500 Hz frequency has a wavelength of 0·6 m; and r.f. energy at 10,000 MHz has a wavelength of 3 cm.

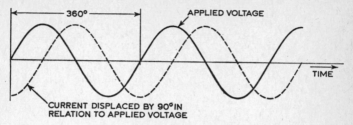

Fig. 8. Phase relationship between current and voltage.

What is phase ?

This term refers to the relationship in time of two alternating quantities. In Fig. 8 the applied voltage and the current are displaced in time and are said to be out of phase. They would be in phase if there were no displacement—that is, if the waves coincided.

In the example the voltage leads the current by 90°. It is also possible for the current to lead. A general rule is that current leads voltage when the e.m.f. is applied to a capacitor (see later), and the current lags when the e.m.f. is applied to an inductor. The amount of lead or lag is called the phase difference.

It should be noted that "phase" can refer to other kinds of wave, such as sound waves; although there are no currents and voltages to compare, the phase relationship of different sound waves may well be of importance in audio engineering.

Are there other types of current ?

In addition to a.c. and d.c. there is current with a pulse waveform. It is of special interest in electronics.

Simple pulsed current will be produced by a circuit (see Fig. 9) in which the source of supply (a battery, say) is repeatedly switched on and off. The current will be d.c., flowing in one direction only; but it will be intermittent—a series of pulses. It has something in common with a.c. in

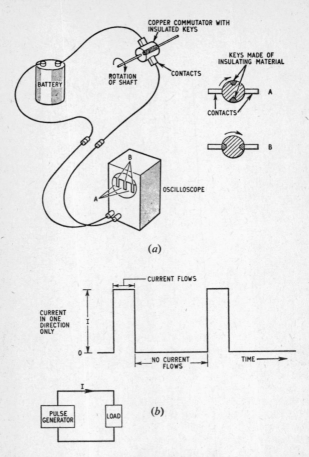

Fig. 9. (a) Producing a pulse waveform by means of a battery and a rotary switch. The load in this circuit consists of an oscilloscope, an instrument which will reproduce on its screen the waveform applied to it, as shown. (b) An ideal pulse waveform.

20

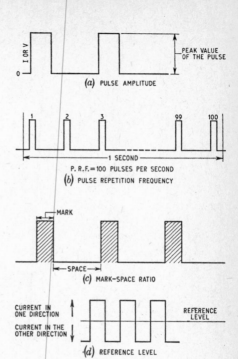

Fig. 10. Illustrating some of the terms used to describe pulse characteristics.

(a) PULSE AMPLITUDE

PEAK VALUE OF THE PULSE

P.R.F.=100 PULSES PER SECOND
(b) PULSE REPETITION FREQUENCY

(c) MARK-SPACE RATIO

(d) REFERENCE LEVEL

that it is a changing quantity, but whereas a.c. goes gradually from minimum to maximum, the pulsed current goes suddenly to its maximum value (the feature which accounts for its usefulness). Pulses, like a.c., can be supplied to a transformer if the conditions are right.

What are the main characteristics of pulses?

We need to know—as a minimum—the following things about a pulse waveform. The value of current or voltage referred to will be the *peak value* (see Fig. 10): the question of r.m.s. value, used for sine waves, no longer arises. The

21

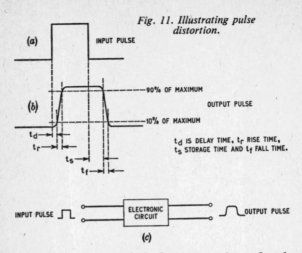

Fig. 11. Illustrating pulse distortion.

(a) INPUT PULSE

(b) OUTPUT PULSE

90% OF MAXIMUM

10% OF MAXIMUM

t_d IS DELAY TIME, t_r RISE TIME, t_s STORAGE TIME AND t_f FALL TIME.

t_d
t_r
t_s
t_f

INPUT PULSE | ELECTRONIC CIRCUIT | OUTPUT PULSE

(c)

next important characteristic is the number of pulses a second (the counterpart of frequency in a.c.), and this is known as the *pulse repetition frequency*.

Pulses last for a certain duration and have certain intervals between them. The pulse length (duration) is termed the *mark* time and the intervals are called the *space* time. This leads us to the *mark/space ratio*, which is of special interest to circuit designers and users. Then there must be a *reference level*, from which the pulse begins and to which it returns. These characteristics are illustrated in Fig. 10.

In practice a pulse will become distorted to some extent: it suffers time lags due to the characteristics of the circuits through which it passes. Such delays lead to the use of four more technical terms—*delay time, rise time, fall time* and *storage time*.

Delay time is the time taken for the pulse to attain 10 per cent of its final maximum value; rise and fall times are the times taken for the output pulse to rise and fall from 10 to 90 per cent of its total value; and storage time is the

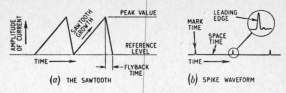

Fig. 12. Two other common types of waveform.

delay caused by the circuit not cutting off immediately when the input pulse ceases. These time lags are indicated in Fig. 11, which shows an input pulse (a square wave) and its appearance as an output pulse—distorted after its passage through a circuit.

What other waveforms are there ?

Among several types of waveform encountered in electronics, a particularly important example is the sawtooth, shown in Fig. 12 (a). There is a gradual increase in value with time, but after the maximum has been reached the current falls in a very short period of time (the flyback time) to the reference level.

Another example is the spike waveform, shown in Fig. 12 (b). This can be regarded as a square wave with a very small mark/space ratio: the main interest is in its initial rise to maximum value.

What is the significance of reactance ?

The size of an alternating or pulse current is determined by the applied voltage and the resistance of the circuit. To this extent the alternating quantity behaves like d.c. But there is another factor affecting the behaviour of an a.c. circuit: it is called reactance and its value is determined by the presence of inductance and capacitance in the circuit.

Inductive and capacitive reactance (both are measured in ohms) limit the amplitude of the current which flows in a circuit when an alternating or pulsed voltage is applied. The value of the reactance also depends on the frequency

23

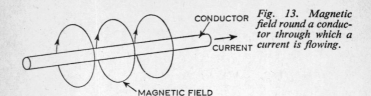

CURRENT

Fig. 13. Magnetic field round a conductor through which a current is flowing.

MAGNETIC FIELD

of the a.c.: inductive reactance increases in value as frequency rises, but capacitive reactance decreases with increase of frequency. Reactance combined with resistance (as in practical circuits) gives *impedance*, again measured in ohms.

What is inductance ?

When current passes along a wire a magnetic field is produced around it (see Fig. 13). The kinetic energy of the current is contained in this field, and therefore the field will tend to prevent a change in the current: it induces an e.m.f. in the circuit in opposition to the change of applied voltage. The effect is known as self-induction and the circuit is said to possess inductance.

A straight wire does not normally provide the amount of inductance required for practical purposes; to increase the inductance the wire may be wound as a solenoid (a coil) on a simple insulated former with a core of air. To obtain a further increase in field intensity and inductance while keeping to a reasonable number of turns of wire, the field

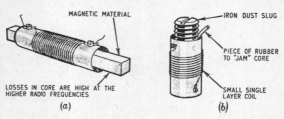

MAGNETIC MATERIAL

IRON DUST SLUG

PIECE OF RUBBER TO "JAM" CORE

SMALL SINGLE LAYER COIL

LOSSES IN CORE ARE HIGH AT THE HIGHER RADIO FREQUENCIES

(a)

(b)

Fig. 14. (a) Inductor with core. (b) Dust-cored inductor.

24

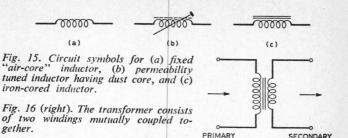

Fig. 15. Circuit symbols for (a) fixed "air-core" inductor, (b) permeability tuned inductor having dust core, and (c) iron-cored inductor.

Fig. 16 (right). The transformer consists of two windings mutually coupled together.

PRIMARY   SECONDARY

can be concentrated by placing a magnetic core inside the former, as in Fig. 14 (a).

A coil—or choke, as it may sometimes be called—for a high-frequency circuit will probably have only a few turns of wire, spaced apart. For audio-frequency work the coil is formed on a core consisting of laminations of, say, nickel-iron. Tuning coils for radio equipment usually have adjustable cores, as shown in Fig. 14 (b), and are said to be permeability tuned. In such cases dust cores are used: they are moulded from a mixture of magnetic particles, insulating powder and a binding agent.

Returning to the characteristics of inductors, it should be noted that the opposition to the flow of a.c. increases as frequency increases. Further, an inductor passes current which is delayed in phase (that is, it lags the voltage) by a quarter of a cycle, or 90°. As already indicated, inductive reactance is measured in ohms.

The relevant circuit symbols are shown in Fig. 15.

What is mutual induction ?

Two coils can be arranged so that they are physically separated but coupled in the electrical sense by interaction between their magnetic fields. Mutual induction is said to occur between the coils. A transformer exploits this effect: it consists of two coils wound on a common core in such a way that there is good coupling between the windings.

A variation of current in one winding produces a

25

proportional change in the flux (magnetic field) in the core, and hence an e.m.f. is induced in the other winding. In practice a varying (usually alternating) voltage is applied to the primary winding and the output is taken from the secondary winding (see Fig. 16). The turns ratio of the transformer's coils determines the relationship between the voltage and current changes on the two sides.

What is the unit of inductance ?

The unit of inductance is the Henry. A device or circuit has an inductance of one Henry if a current, varying at the rate of one ampere per second, induces a back-e.m.f. (the opposition voltage) of one volt.

Practical inductors have values expressed in millihenries (mH) and microhenries (μH). Inductance is denoted by the symbol L.

How does a capacitor work ?

A very simple capacitor consists of a pair of closely spaced metal plates separated by a "dielectric"—an insulator such as air, paper or ceramic. If a battery is connected to this device an electric field will be set up between the plates. Some electrons will flow from the negative battery terminal to the corresponding plate, and a further quantity of electrons will flow from the other plate to the battery's positive terminal. Thus it looks as if the current flows round the circuit, including the dielectric. Actually, however, the electrons which reach the first (negative) plate accumulate there, building up a negative charge; while the electrons

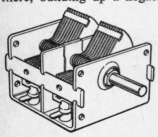

Fig. 17. Two variable tuning capacitors "ganged" together on a common spindle. The air between the plates is the dielectric. Capacitors of the type shown are commonly used for tuning the input circuits of radio receivers.

26

which flow from the other plate give this a positive charge. The result is a p.d. across the plates.

This p.d. forms gradually (relatively speaking, of course) due to the presence of resistance (of the wires, battery and so on) in the circuit. The charge stored by the capacitor depends on the size and separation of the plates, the dielectric, the applied e.m.f. and the time for which the current flows.

The foregoing applies to the capacitor in a d.c. circuit. In the case of a.c. the capacitor appears to conduct (because of the charging and discharging of the capacitor by the a.c.), but the current is advanced in phase: that is, the current leads the voltage by a quarter of a cycle. (Compare this with the inductor previously mentioned.) Thus a capacitor blocks a d.c. voltage but not an a.c. voltage. Like inductive reactance, capacitive reactance is measured in ohms. The higher the frequency and the bigger the capacitance the smaller the reactance.

What is the unit of capacitance ?

The capacitance in Farads (named after Faraday) is the number of coulombs (quantity) of electricity entering the capacitor as a result of an e.m.f. of one volt.

Practical capacitors encountered in electronics have values expressed in microfarads (millionths of a Farad, abbreviation μF) and picofarads (millionths of a microfarad, abbreviation pF). Capacitance is denoted by the symbol C.

Describe some practical capacitors.

Capacitors with an air dielectric are commonly used for the adjustment of tuned circuits in radio equipment. A number of moving vanes are ganged together on a spindle in such a way that they can be moved in relation to another fixed set of vanes (see Fig. 17), thus varying the capacitance over a certain range. Many fixed capacitors, with relatively high capacitances (100 μF or more), have spirally wound metal foil plates with a paper dielectric, as shown in Fig. 18 (a). Small fixed capacitors, often with values of only a

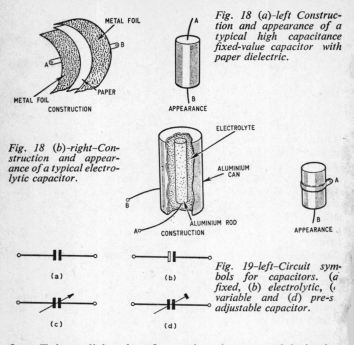

Fig. 18 (a)–left Construction and appearance of a typical high capacitance fixed-value capacitor with paper dielectric.

Fig. 18 (b)–right–Construction and appearance of a typical electrolytic capacitor.

Fig. 19–left–Circuit symbols for capacitors. (a) fixed, (b) electrolytic, (c) variable and (d) pre-set adjustable capacitor.

few pF, have dielectrics of ceramic, mica or special plastic. Electrolytic capacitors offer the advantage of very high capacitance for small physical size. In this type a thin oxide film—an insulator—is formed on a metal strip by electrolysis and becomes the dielectric (see Fig. 18 (b)).

The electrolytic, which only works with one polarity of applied e.m.f., is frequently used where it is required to smooth out ripples on direct current. It is also commonly found in transistor equipment. Other types of capacitor are commonly used as "blocking" components: they block d.c. but allow the alternating component to pass on to the next stage in the circuit.

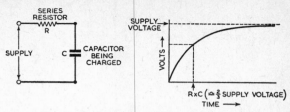

Fig. 20. The time constant of an RC circuit.

Quite apart from the components mentioned here, the question of unwanted capacitance can also arise. For example, a small capacitance exists between wires and connections in circuits and between the electrodes in valves. These are called *stray capacitances* and may interfere with the action of some types of circuit. The problem of avoiding them is one of the design and layout of components and equipment; the solution becomes more difficult to achieve as frequency is increased.

Some circuit symbols are shown in Fig. 19.

What is meant by "time constant"?

With a circuit consisting of resistance and capacitance as shown in Fig. 20, the capacitor will not fully charge immediately a supply voltage is applied to the circuit. As it charges, the voltage across the resistor will fall and so the charging current, flowing through the resistor, will decrease, in accordance with Ohms's Law. The time constant of such a circuit is given by the formula $RC = T$, where R is in ohms, C is farads and T in seconds, and in practice is the time taken for the voltage across the capacitor to reach approximately $\frac{2}{3}$ of the voltage applied to the circuit. Thus the time constant of an RC combination determines the speed at which the circuit responds to signals applied to it, an important point in many applications. Similar effects occur with circuits consisting of resistance and inductance. In this case the time constant is the time taken for the current through the inductor to reach $\frac{2}{3}$ of its maximum

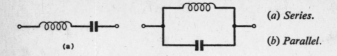

(a) Series.

(b) Parallel.

Fig. 21. Resonant tuned circuits.

value after the application of an e.m.f., and is given by R/L with L in henries and R in ohms.

What is the effect of combining inductance and capacitance ?

A *resonant circuit* can be formed from a combination of inductors and capacitors, as shown in Fig. 21. A circuit of this kind responds more strongly to a particular frequency, and this property of frequency selection is of great importance in radiocommunications and other branches of electronics.

At some frequency, depending on the values of the components used, the condition of *resonance* will occur. In the series circuit, at frequencies other than the resonant frequency, there is impedance to the flow of current due to the reactance of the components. At resonance the inductive and capacitive reactances are equal and cancel out, and the opposition to current is limited to the residual or "equivalent" resistance (r) of the inductor. Thus the series circuit

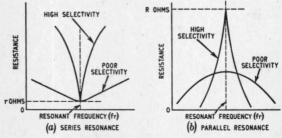

Fig. 22. Illustrating the selectivity of tuned circuits.

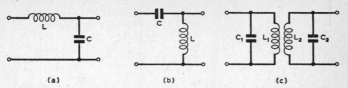

Fig. 23. Filter circuits. (a) low-pass; (b) high-pass; (c) band-pass.

passes currents at its resonant frequency (to which it is said to be *tuned*); or it may be used to short-circuit unwanted currents at this frequency. The parallel circuit, however, offers maximum impedance—it is the equivalent of a large resistor—at its resonant frequency. Current flow is then at the minimum but the voltage across it at the maximum. It will thus block alternating currents but will pass an a.c. voltage at this frequency.

The sharpness of this discrimination—called the *selectivity*—of such circuits is of special interest and depends on certain design factors. The Q of the circuit is especially important (the Q factor of the inductor, a measure of its "goodness", is the ratio of its reactance to its resistance). Selectivity is illustrated in Fig. 22. In some applications we may require the circuit to respond over a wide band of frequencies rather than give good selectivity. We then deliberately reduce the Q of the circuit. This is done by adding external resistance to the circuit as appropriate, and is called *damping*.

Inductors and capacitors can also be combined so as to form filters which *attenuate* (cut) all frequencies above or below a certain point. The arrangement shown in Fig. 23 (a) is a low-pass filter: it passes current up to the series resonant frequency and attenuates frequencies above this. Interchanging L and C makes the circuit a high-pass filter, shown in Fig. 23 (b). A band-pass filter, shown in Fig. 23 (c), allows a narrow band of frequencies to pass and attenuates the bands on either side. In its common form, as used for example in radio receivers, this filter comprises a transformer with tuning capacitors connected across the primary and secondary windings.

What is a signal ?

Most electronic equipment exists for the purpose of producing, detecting or in some way handling a signal, which may take the form of an oscillating current, a varying direct current, a series of pulses or one of a number of other electrical quantities.

As in fields not connected with electronics, the signal may convey information or initiate action. We talk of the radio signal which is received via an aerial, and of a test signal which is generated by one piece of equipment and injected into another for the purpose of measurement or fault-finding. We may simply use the signal without being concerned with what happens to it as it passes through a circuit (if, for example, we require only to confirm its presence); or we may be preoccupied with the problem of preventing distortion of the signal as it passes from the input to the output terminals of an amplifier.

What is meant by "noise" ?

Obviously, a signal is generated and used with some care because it is wanted for a specific purpose. However, in practice there are always other, spurious signals which tend to interfere with the wanted signals, and these are generally referred to as *noise*.

Noise may be created by machines or other equipment, and then induced into circuits carrying the wanted signals. However, designers and users of electronic equipment are more concerned with noise of a fundamental kind: that due to the random movement of electrons in conductors or in valves and other components.

An important design objective is a good *signal-to-noise ratio* (a self-explanatory term), though noise cannot be entirely eliminated. A sensitive and relatively noise-free amplifier (valve or transistor) in the first stage of an equipment is an outstanding requirement if noise at its output is

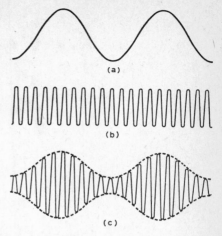

Fig. 24. Modulation: (a) audio signal; (b) carrier wave; (c) carrier wave modulated by the audio signal. This is amplitude modulation. With frequency modulation the audio signal modulates (varies) the frequency of the carrier wave signal.

(a)

(b)

(c)

to be minimised (since any noise in the first stage is amplified in each succeeding stage).

Why is it necessary to "modulate" a signal ?

A signal is modulated so that it can be transmitted over a considerable distance. The most familiar example is in radio, where it is required to broadcast widely an audio signal, derived from sounds (which, obviously, cannot themselves carry very far). Before any audio signal is applied to it, the radio transmitter sends out a simple, unvarying signal in the form of an electromagnetic wave at some high frequency. This is known as the *carrier wave*.

When information is broadcast, the audio signal is made to vary the strength of this carrier wave, so that the latter conveys the signal. The modulation of a carrier wave by an audio wave is illustrated in Fig. 24. Therefore, in general, modulation involves alterations in the amplitude (or it may be frequency) of an oscillation by a signal of some lower frequency. In many electronic applications today various forms of pulse modulation are used. In these information is

33

conveyed by varying the heights, widths, positions, etc. of a train of pulses.

How do electromagnetic and sound waves compare ?

Electromagnetic waves embrace a range of wave-motions which do not require a material medium for their propagation. Examples are x-rays, ultra-violet rays, visible light, infra-red rays and radio waves (in order of increasing wavelength). They travel with a velocity of 300,000 kilometres a second.

Sounds are transmitted as longitudinal pressure waves, and they require a material medium (air, liquids, solids, etc.) for their propagation. Their velocity depends on the propagating medium and on temperature. In air at 0°C the velocity is about 341 metres per second.

What are ultrasonic waves ?

These are pressure waves, similar to sound waves but at frequencies above the limit of audibility (that is, above about 20 kHz). Ultrasonic energy is increasingly employed in industry, medicine and scientific investigations, and the potential applications appear to be almost unlimited. Industrial and other branches of ultrasonic technology have stemmed from undersea echo sounding, signalling and related applications.

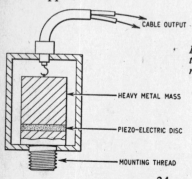

CABLE OUTPUT

Fig. 25 (a): A piezo-electric transducer for the measurement of acceleration.

HEAVY METAL MASS

PIEZO-ELECTRIC DISC

MOUNTING THREAD

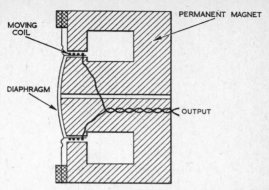

Fig. 25 (b). The moving-coil microphone. Sound waves impinge on the diaphragm, causing the coil to move in the gap of the permanent magnet, resulting in electrical signals at the output.

What is a transducer?

A transducer is a power transforming device—one which converts energy into some other form. For instance, a microphone converts sounds into an electrical signal, and a gramophone pick-up converts mechanical movements into electrical signals. Other devices, notably those used in industrial electronics, have such tasks as sensing temperature, acceleration or pressure, producing corresponding electrical signals for purposes of measurement. A transducer may also convert electrical energy into mechanical energy.

Electromagnetic, photo-electric, piezo-electric or other principles may be employed. Piezo-electric transducers are often relatively inexpensive and are available in great variety. A suitable material, for example lead zirconate titanate, will produce an e.m.f. when it is subjected to mechanical force, such as bending or twisting. Conversely, the material will exert force if electrical energy is applied to it. Fig. 25 shows a piezo-electric transducer designed for the measurement of acceleration (a), and a moving-coil microphone (b).

3

VALVES, TRANSISTORS AND OTHER COMPONENTS

What is thermionic emission ?

The release of electrons from materials under the action of heat. Reference has already been made to the presence in some materials of "free electrons", which are loosely bound in the atomic structure and therefore able to drift away when suitably influenced. To produce free electrons in a more controlled manner—in valves, for example—heat may be used. Heating a material—a metal, say—will cause an increase in kinetic energy among the atoms and electrons, with consequent collisions between the particles. The result is that some of the free electrons are given enough energy to escape from the metal, as indicated in Fig. 26. If the heated metal is exposed, the electrons will be collected by the surrounding air.

Fig. 26. Thermionic emission: free electrons escape from the heated metal.

How is this effect employed in valves ?

In the case of a valve a metal plate, or electrode, is placed inside a glass or metal envelope. Usually there is a vacuum in the envelope (although for special purposes a gas may be introduced: see later). The electrons can not be collected

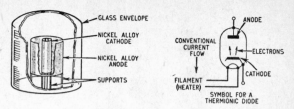

Fig. 27. The thermionic diode valve: construction and symbol.

by a vacuum and therefore they form a cloud—called a
space charge. In practice a dense cloud is required, and a
metal with suitable properties is chosen for the emitting
electrode—the *cathode*, as it is called. Cathodes of oxide-
coated nickel alloy are generally used since they are rich
in free electrons.

Although the cathode could be a filament heated by an
electric current passing through it (as in a lamp), the more
usual arrangement in valves is a separate tungsten wire
filament placed near to the cathode. The filament is then
merely a heater and is electrically isolated from the cathode.

What is a diode valve ?

The diode has two electrodes—cathode and *anode*. The
arrangement is shown in Fig. 27 together with the circuit
symbol. The heated filament raises the temperature of the
cathode, which emits a cloud of electrons, and these remain
near the cathode to form the space charge. If the anode is
made positive with respect to the cathode by applying a
voltage V_a (see Fig. 28), a current I_a flows through the

*Fig. 28. Current
flow through a
diode valve.*

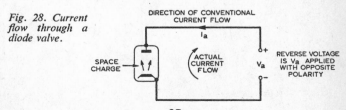

valve: the current consists of electrons moving from the space charge to the anode.

Note that the electron movement is from cathode to anode (from negative to positive). However, in early work on electrical phenomena it was thought that currents flowed from positive to negative; this convention is still used in some fields, and current assumed to be flowing from positive to negative is called *conventional current flow*. The point is illustrated in Figs. 27 and 28.

The diode conducts in one direction only—from cathode to anode, as described above. It will not conduct in the opposite direction. Therefore it behaves like a switch, offering a low resistance in one direction. Prominent among its many uses in electronics are rectification and detection. It is a non-linear device.

What is meant by the terms linear and non-linear ?

Linear components obey Ohm's law, which states that the ratio of the voltage to the current in a circuit is constant. This ratio is the resistance of the circuit. (As already indicated, "impedance" is used instead in an a.c. circuit where inductance or capacitance is also present.) The linear relationship is easily visualised: if the voltage is doubled the current will double.

However, numerous components used in electronics, including the diode and other types of valve, do not obey this simple law. An increase in voltage does not produce a linear increase of current, and therefore such components are said

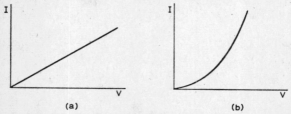

Fig. 29. Voltage/current relationships: (a) linear, (b) non-linear.

to be non-linear. Linear and non-linear characteristics are illustrated in Fig. 29.

What is a semiconductor diode ?

Semiconductor devices are displacing thermionic diodes and other valves from many of their applications for a variety of reasons, both technical and economic. For instance, semiconductor diodes have such advantages as absence of heater supplies, extremely small size and weight, good reliability and small self-capacitance. (Thermionic diodes and other valves still show marked advantages in certain applications however.)

The elements used most widely in the manufacture of semiconductor devices are germanium and silicon, although some other materials are now assuming importance. The addition to these elements of very small amounts of impurities, known as doping agents, provide the required electrical characteristics. For example introducing arsenic into the semiconductor produces what is called an n-type material, which contributes electrons to the device's action; and indium produces a p-type material, which can contribute holes. As mentioned in Chapter 1, holes are spaces left by the loss of electrons: they are opposite in sign to electrons and can accept electrons. A semiconductor diode consists of a p- and an n-type semiconductor formed together to make a pn junction.

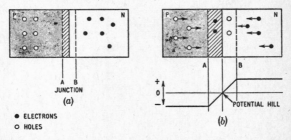

Fig. 30. (a) Semiconductor pn junction; (b) potential barrier.

39

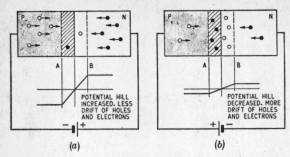

Fig. 31. Biasing a pn junction. (a) Reverse biasing: potential barrier increased; (b) forward biasing: potential barrier reduced.

How does a semiconductor diode operate?

Fig. 30 shows a junction diode formed by p-type and n-type material. Some of the electrons and some of the holes possess sufficient energy to diffuse across the junction, forming a *depletion layer* on either side by the neutralisation of holes in the p region and free electrons in the n region. The part of the n-region nearest the junction is thus given a positive charge and the part of the p-region nearest the junction a negative charge. This diffusion of holes and electrons continues until a *potential barrier* or *hill* is formed; this then prevents further movement of holes and electrons. The action is a little like charging a capacitor.

The barrier can be reduced or increased by applying a voltage. In Fig. 31 (a) the applied e.m.f. (from the battery) is shown *reverse biasing* the junction: the potential barrier is increased and the diffusion of electrons and holes, i.e. the flow of current, is reduced. Alternatively the junction may be *forward* biased as shown in Fig. 31 (b): the potential barrier in this condition is reduced and the diffusion of electrons and holes is increased (that is, a current flows).

Thus, like the thermionic diode, the semiconductor diode allows unidirectional current flow. Again, it is a non-linear device. Fig. 32 shows the circuit symbol: note that it points in the direction of "conventional current" flow and that electron flow is in the opposite direction. (This symbol is also used to denote a metal rectifier, which is similar in action to a semiconductor diode but based on the properties of certain metals instead: it is a more bulky device than its semiconductor equivalent.)

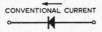

CONVENTIONAL CURRENT

Fig. 32. Circuit symbol for a semiconductor diode or a metal rectifier.

How is a diode used for rectification ?

Since the diode permits current flow in one direction only, it can be used in an a.c. circuit for the purpose of rectification (the conversion of a.c. to d.c.). Current will flow only during the half-cycles of the a.c. waveform when the "anode" of the diode is positive with respect to the "cathode": the device conducts on the positive half-cycles and the output is a series of positive pulses which can subsequently be smoothed out to give a steady direct current.

Rectifiers can be connected in various ways according to the nature of the *load* (the circuit to which the current is being supplied) and the performance required. Two simple arrangements are shown in Fig. 33. In the first of these, current from the transformer secondary flows through the rectifier and into the load on alternate (positive) half-cycles (actually the rectifier acts rather like a switch: on positive

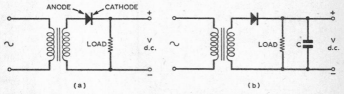

(a)   (b)

*Fig. 33. Two simple diode rectifier circuits: (a) has a resistance load;
(b) has a capacitor added to the load to reduce the "ripple" in the output.
These circuits provide "half-wave" rectification.*

41

half-cycles it "opens", the current then flowing round the circuit via the load and the rectifier).

In the second arrangement a capacitor is added for the purpose of smoothing the "ripple" left on the d.c.: it charges when the rectifier conducts and discharges when the rectifier is not conducting, thereby smoothing the d.c. output of the rectifier.

Since the ripple voltage across the capacitor is equal to the amount the capacitor voltage falls during its discharge period, a reduction in ripple must follow from increasing the value of the capacitor.

More elaborate circuits are frequently used in order to achieve full-wave (instead of half-wave) rectification, and it is possible to connect rectifiers so that they produce a greater d.c. voltage than the a.c. input voltage.

How do valves amplify?

So far only the diode, useful for rectifying and switching, has been mentioned. For amplification of signals the simplest valve that can be used is the triode, in which a third electrode, the control grid, is added between the anode and cathode. This is shown in Fig. 34 together with the circuit symbol.

The control grid has a mesh structure so that the electrons emitted by the cathode can pass through it. In fact some electrons collect on it, producing a negative charge, but these can be allowed to return to the cathode externally via a "grid leak" resistor (see Fig. 35).

The control grid is so named because it controls the flow of current in the valve. If the grid is made slightly negative with respect to the cathode an electrostatic field will be created around it, tending to repel electrons which are moving towards the anode. Electrons which escape this influence pass through the grid to form the anode-cathode current. If the grid is made sufficiently negative, the electron flow to the anode will cease, and the valve will be in what is termed the *cut-off* condition. Conversely, when the grid is slightly positive maximum current flows through the

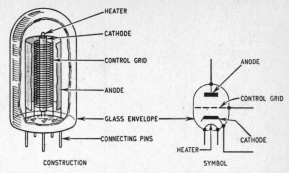

Fig. 34. Thermionic triode valve: construction and symbol.

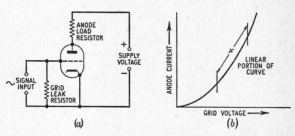

Fig. 35. (a) Simple triode amplifier circuit. (b) Characteristic curve. Bias is applied to the valve so that it operates over the straight portion —X— of the curve.

valve: in this condition the control grid no longer determines the current flow and the valve is said to be *saturated* or *bottomed*.

Thus a negative grid voltage controls the anode current, and in fact a very small signal voltage—a millivolt, say—will serve to control a substantial current. The grid does not have to supply a current; indeed, it may be that no significant current is available—as when a microphone is connected to the input of an amplifier.

43

In a practical circuit a *load resistor* of high value is inserted in the valve's anode connection, as shown in Fig. 35 (a). Since a varying voltage at the grid is varying the current flowing round the circuit, there will be a varying voltage across this resistor. This voltage variation will be much greater than the signal's variation, and therefore the valve will function as an amplifier.

Note that as connected in Fig. 35 (a) there will be phase reversal between the input and output (the voltage across the anode load will be increasing—and the voltage *at* the anode decreasing [measured to earth]—as the grid voltage increases). Occasionally, especially in pulse circuits, it is this feature rather than amplification that is required. A valve or transistor used in this way is called an *inverter*.

What is meant by a triode's characteristic ?

A characteristic curve is a plot of grid voltage and anode current. As Fig. 35 (b) shows, the grid voltage and anode current do not have a linear relationship: valves are non-linear, as stated earlier. This means that the voltage developed across the load resistor, though amplified, may not be a replica of the grid voltage. In other words, it may be distorted. Fortunately it is possible for the relationship to be made linear over a part of the characteristic, as shown in Fig. 35, and provided the valve is operated correctly distortion can be reduced to a very small amount. Correct *biasing* is the key to this.

The mid-point of the linear portion of the characteristic is called the *bias point*, and the grid of the valve can be supplied with a fixed negative voltage so as to ensure that the a.c. signal variations are centred on this point, operation of the valve then being over the linear portion of the curve provided that the signal does not swing too far.

If a resistor is inserted in the cathode lead, a voltage will be developed across it and the cathode will be made positive with respect to earth. Then, with the grid connected to earth via the grid leak, there will be a p.d. between cathode and grid, and the grid will be negative with respect to the

44

cathode. The value of the resistor is chosen so that the grid voltage is at the bias point. This is the method generally used in amplifiers, but a capacitor is often added across the cathode bias resistor, to prevent the bias varying as the signal current varies.

What is the gain of an amplifier ?

The degree of amplification is referred to as the gain, and this is expressed in decibels (abbreviation dB). The decibel is used for the ratio of two quantities (voltages, powers, sounds, etc.), and with amplifiers the ratio is that of the output voltage to the input voltage (note, however, that the input and output impedances must be the same for this to hold true).

Values of gain, expressed in dB, can be added together. Thus the output of an amplifier with a gain of 40 dB could be connected to the input of an amplifier with a gain of 60 dB: the total gain would be 100 dB.

Why are tetrode and pentode valves used ?

Although the triode continues to be used successfully in a number of applications, it presents some disadvantages that more elaborate valves are designed to overcome. For example, there is the build-up of a space charge which impedes electron flow; and the use of a fine mesh grid, intended to provide improved control over electron flow, leads to increased grid area and a greater inter-electrode capacitance.

A second grid—the screen grid—can be inserted between the control grid and anode to act as an electrostatic shield and reduce capacitative coupling. This type of valve is a tetrode (four electrodes in all). By making the extra grid positive, electrons will be attracted more efficiently from the cathode, helping to overcome the space charge effect. Amplification can be increased. The tetrode is shown in Fig. 36 (a).

The characteristic of the tetrode is not altogether suitable for purposes of amplification. It can be improved by inserting

45

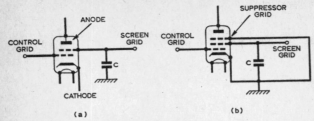

Fig. 36. Symbols for (a) tetrode and (b) pentode valves.

a further grid—called the suppressor grid—between the screen grid and anode. Usually this extra grid is connected to the cathode, as shown in Fig. 36 (b), and in this way it is made negative with respect to the anode. A five electrode valve is called a pentode. An alternative approach is to fit "beam-forming" plates between the screen grid and anode: such a valve is called a beam tetrode.

What is an electrometer valve?

This is a specially constructed valve, with very high insulation resistance (between electrodes) and input impedance, used where it is desired to detect and measure very small currents of the order of 10^{-15} amps or less. The most important requirement of such a valve is that its grid current must be very small for it to be sensitive to such minute inputs. The main applications are for amplifying the output from radiation detectors and in specialised meters.

Which types of valve are gas-filled?

Although most valves are evacuated—the action takes place in a vacuum—some special-purpose devices are filled with argon, helium or neon. An example is the gas-filled triode, or thyratron, used as a switching component. There is also a tetrode thyratron, incorporating a second grid. A gas-filled valve is called a "soft" valve, whereas an evacuated valve is "hard".

With the grid of the thyratron held negative with respect

to the cathode there is no current flow between anode and cathode. However, if the grid is suddenly made positive by the arrival of a positive waveform (e.g. a pulse) the thyratron starts to conduct. The electrons moving through the thyratron collide with the gas particles, resulting in a big increase of free electrons and a large increase in anode current, which reaches a maximum value determined by the applied voltages and the resistance in the anode circuit. Once this has happened the current can no longer be controlled by the grid: it continues to flow until the anode voltage is reduced below the valve's "striking" value. Thus the control grid of a thyratron merely serves to switch it on. These characteristics lead to the thyratron's use in high-speed switching. It can control heavy currents in this purely electronic fashion, and the sparking which would occur in a mechanical switch can therefore be avoided.

Another well-known device is the neon tube, filled with either pure neon or a mixture of neon and argon. It is like a diode, with two electrodes—anode and cathode. With a low voltage applied across the electrodes there is no current flow, but as the voltage is increased a "striking" value is reached. Then current begins to flow and the movement of electrons produces collisions with the gas molecules, causing ionisation. Consequently there is a rapid increase of current and the resistance between anode and cathode falls to almost zero. The ionised gas is seen to glow. The neon tube is used as an indicating device and as a voltage stabiliser.

How does a transistor function ?

The semiconductor diode has been described as a junction of p-type and n-type semiconductor material (germanium or silicon). The simplest type of transistor consists of two such junctions: there are three "layers" of semiconductor, and these can be in p-n-p or n-p-n form.

There are three connections, to the emitter, base and collector, and to some extent the device corresponds to the triode valve. The p-n-p arrangement is represented diagramatically and by a circuit symbol in Fig. 37. The base

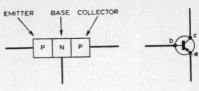

Fig. 37. Diagrammatic arrangement (left) of a pnp transistor and the circuit symbol (right). NPN transistors are denoted by reversing the direction of the emitter arrowhead. They require power and bias supplies of opposite polarity to pnp transistors.

(n-type) region is thin and lightly doped (e.g. with antimony) to contribute electron charge carriers. The regions on either side are heavily doped with indium to give holes.

There are three basic transistor circuit arrangements—the common (or grounded) emitter, common base and common collector configurations (see Fig. 38).

The three configurations have the characteristics listed in Table 3.

In the common base arrangement of Fig. 39 the emitter-base circuit is forward biased and the collector-base circuit reverse biased. A change in emitter current produces a similar change in the collector-base current due to an

Table 3. Transistor circuit characteristics

Configuration	Common Emitter	Common Base	Common Collector
Input impedance	medium	low	high
Output impedance	medium	high	low
Current gain	high	less than unity	high
Voltage gain	high	high	less than unity
Phase reversal	Yes	No	No

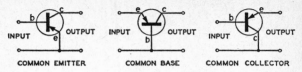

Fig. 38. The three basic ways of connecting a transistor in circuit.

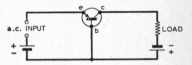

Fig. 39 (right). A pnp transistor in the common base connection, showing bias and supply arrangements.

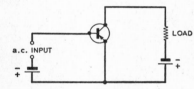

Fig. 40 (left). A pnp transistor in the common emitter connection. Bias and supply polarities for the common collector connection follow this, but the load is placed in the emitter lead.

increased diffusion of electrons or holes across the junctions. The impedance of the two circuits is different and thus some power gain can be achieved (since a current variation in the low resistance emitter-base circuit has become a similar current variation in the high resistance collector-base circuit).

In the more generally used common emitter arrangement shown in Fig. 40 there is current gain between the input and output sides; and since the input resistance is again less than the output resistance (though not nearly as much as in the common-base circuit) there will be high voltage and power gains.

Describe some other semiconductor devices.

A particularly important example is the silicon controlled rectifier, which is similar in many respects to the thyratron. Two p-type sections and two n-type sections of silicon are

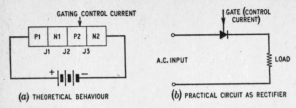

| P1 | N1 | P2 | N2 |

J1   J2   J3

+ |·|·|· −

(a) THEORETICAL BEHAVIOUR

GATE (CONTROL CURRENT)

A.C. INPUT   LOAD

(b) PRACTICAL CIRCUIT AS RECTIFIER

Fig. 41. The silicon controlled rectifier or thyristor. (a) Arrangement of the four p and n sections, giving three junctions. (b) Connected in circuit as a rectifier. With an a.c. input the thyristor will not conduct until a pulse is applied to the gate: it will then conduct, the output following the sinusoidal voltage of the input. The mean d.c. output is thus controlled by the timing of the arrival of the triggering pulses at the gate.

arranged as shown in Fig. 41. There is a connection known as a *gate* electrode in addition to the end connections.

The device presents a high impedance to current flow until either a certain critical voltage is reached (across it) or a triggering pulse is applied to the gate. Conduction is then started, but the gate loses control, and the low impedance condition remains until the supply voltage is removed or a reverse voltage applied.

Another useful device is the zener diode. If a semiconductor diode is reverse biased, very little current flows. A sufficiently large reverse voltage however causes zener breakdown, when the current suddenly jumps to a high value which is determined by the resistance in the circuit. At this point changes in voltage have no effect on the current flow. The zener diode is designed to make use of this characteristic, the device being valuable for voltage stabilising purposes.

What is a photo-electric cell ?

This is a device for converting light energy into electrical energy. Selenium may be used as an active element: light reaches it (after passing through an outer oxide layer) and liberates photoelectrons which form an electric current.

The current varies in proportion to the amount of light. Thus the cell is of value as a detecting device, particularly in industrial plant.

The phototransistor should also be noted. This is a junction transistor with a light-sensitive base, and it is used as an amplifier to operate, via a relay, a variety of types of equipment (such as machinery controls and burglar alarms). A phototransistor can be designed to work with an infra-red source instead of with visible light.

In this context it should be mentioned that certain semi-conductors can be made to *produce* light when passing an electric current. Gallium phosphide is a particularly promising material.

What is a thermistor ?

A thermistor enables heat to be changed into electrical energy. It is a temperature-sensitive resistor in semiconductor form. The device has a negative temperature coefficient: that is to say, its resistance decreases as temperature increases. This resistance change can be detected and used in, for example, a measuring circuit, or where it is required to correct a circuit's behaviour when the temperature fluctuates.

4

AMPLIFICATION

What are the main features of a voltage amplifier?

The main components of a simple triode amplifier are shown in Fig. 42. As outlined in the previous chapter, the signal, applied to the control grid, causes a variation in the electron flow from cathode to anode, and this varying current flows through the anode load resistor.

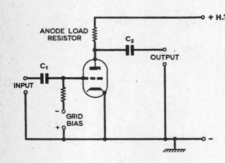

Fig. 42. A simple triode valve voltage amplifier circuit. C_1 and C_2 are coupling capacitors.

The p.d. developed across this resistor is proportional to the anode current and to the resistance. It is much greater than the grid voltage but follows the input waveform. However, the anode and grid potentials change in opposite directions (that is, they have opposite polarity at any instant).

In order to obtain sufficient amplification the load resistor is made high in value, but there is some upper limit which is determined by the valve's characteristics. The idea of bias was introduced earlier: it was noted that cathode biasing could be obtained by inserting a resistor in the cathode lead. In Fig. 42 a separate source of bias voltage is indicated.

Two coupling capacitors are also shown. Since the amplifier is dealing with alternating—or at any rate rapidly fluctuating—voltages it is possible to feed in the signal via a capacitor and also to extract the output from the anode in similar fashion. The purpose of the capacitors is to isolate the high-tension supply at the anode of the valve from the grid of the stage that follows it. (As stated earlier, a capacitor blocks d.c. and passes a.c.) Alternatively in some applications transformers are used for this purpose: they also block d.c. but pass a.c.

Valve amplifiers of the kind shown here can be connected together to give more amplification than can be obtained with one stage. *Cascade* connection is used: the valves are connected in series, with the output of one fed into the input of the next. Only a limited number of stages can be connected in cascade, however, as increasing the number increases the possibility of some of the amplified signal leaking back to the input, resulting in violent instability.

How does a simple transistor amplifier operate ?

The behaviour of a simple transistor amplifier is somewhat similar to that of a triode amplifier. A changing base potential causes a corresponding variation in the flow of electrons from emitter to collector, and this proportionally varies the collector current. If the collector load resistor is large enough the resultant change of collector potential is much greater than the changing base potential.

Supply and bias current polarities depend on whether a p-n-p or n-p-n transistor is used and also on the circuit configuration. A widely used arrangement is a p-n-p transistor in a common emitter circuit, as shown in Fig. 43 (a). Negative bias, to make the transistor operate at the correct point on its characteristic, is obtained via the resistor R

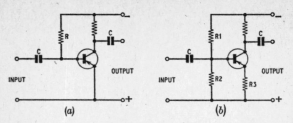

Fig. 43. Simple common emitter pnp transistor amplifiers, showing two different biasing techniques.

from the supply. As with the valve amplifier a coupling capacitor *C* is needed. An n-p-n transistor would function in a similar circuit but the supply voltage would be of opposite polarity.

The transistor is sensitive to temperature, and the amplifier's performance will be seriously affected by temperature changes unless something is done to stabilise it. The simplest method is shown in Fig. 43 (b): suitable resistors are connected between emitter and earth and between base and earth. Thermistors, mentioned in the previous chapter, are also sometimes used.

Finally, the most important differences between transistor and valve amplifiers (as far as circuit operation is concerned) should be noted. As already stated, the valve's anode current is controlled by grid voltage variations, and this will happen without a flow of grid current. With the transistor, on the other hand, a basic requirement is a flow of current in the "control" section—the emitter-base circuit in the common emitter or common base configurations— and it is variations of this current that provide control of the collector current. Another big difference is the low input impedance of the transistor circuit, compared with the equivalent valve amplifier. In addition the transistor works with a low supply voltage, 6 or 12 volts for example: in contrast the valve needs an h.t. supply, usually of about 200 volts.

How are amplifiers classified ?

Valve amplifiers are classified according to the conditions determining the flow of current through the valve—the setting of the grid bias and the value of a.c. volts applied to the control grid. A similar classification is used for transistor amplifiers.

Most voltage amplifiers operate in class A, and this mode may also be used for power amplification. The grid bias and input voltages are such that anode current flows at all times. With class B operation the bias is about equal to the cut-off value, so that the anode current is approximately zero when no grid voltage is applied. Current flows through the valve for approximately half of each cycle while an a.c. grid voltage is applied. There is also class C and such intermediate classes as AB_1 and AB_2.

How do output stages differ from voltage amplifiers ?

In, say, an audio amplifier, a number of stages of voltage amplification are followed by an output stage which must deliver power. Therefore the valve or transistor used in this stage must be suitable for a heavy flow of current, and the load must be connected in such a way that maximum power transfer is obtained.

It is usual to match a valve output stage to its load (e.g. loudspeaker) by means of a transformer, as shown in

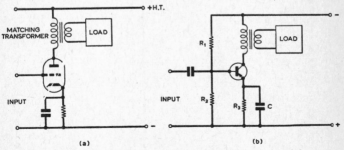

Fig. 44. (a) Valve and (b) transistor used as a power amplifier.

Fig. 44 (a). A transformer may also be used in a transistor amplifier, as shown in Fig. 44 (b), but it is not difficult to devise a circuit in which reasonably good matching is secured without the intervening transformer.

What kinds of distortion are encountered ?

Non-linearity is a form of distortion that is encountered in all amplifiers. As the input increases, the output should increase in corresponding fashion; if it does not, the amplification is *non-linear*. In order to minimise distortion the components must be operated within certain limits. Careful design and appropriate expenditure on good components and techniques can result in very low distortion levels. Another common form of distortion, *frequency distortion*, is due to variation of amplifier gain with frequency. Other troublesome forms of distortion are *intermodulation distortion*, where different parts of a complex signal interact and distort one another, and *harmonic distortion*, caused by harmonics of the signal being handled being spontaneously generated and therefore present in the output.

What is push-pull amplification?

Push-pull connection of a pair of output valves is used to increase output power. The basic arrangement is shown in Fig. 45. Signals of opposite phase are applied by an

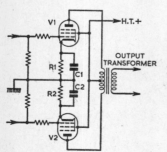

Fig. 45. Basic valve push-pull output stage. Further improvement in performance is possible by obtaining the screen supplies from tappings on the output transformer primary. This introduces feedback (see later) within the stage.

earlier "phase-splitting" stage to the control grids of the output valves. As the grid voltage of one valve becomes more negative the grid voltage of the other becomes more positive; and since the grids are in opposite phase the anode current in one valve increases while that in the other decreases.

The valves are matched to their load by a centre-tapped transformer the primary of which is connected between the anodes of the two valves. Changes in the currents through the valves produce additive effects in the transformer secondary winding: this is because the currents pass through the two halves of the primary winding in opposite directions so that they assist one another. Push-pull connection can be used to minimise distortion while increasing output, and variants of the basic arrangement are widely used to secure further improvements in efficiency and distortion reduction.

Transistors are also used in push-pull stages—often in the class B mode. This is a very efficient arrangement because the total power dissipated in the output stage falls to a low figure during periods of no signal input. In such circuits care has to be taken to minimise an effect known as *crossover distortion*, which can occur as one transistor of the pair takes over from the other. Special class AB modes are increasingly used where high standards of performance are required.

What is meant by bandwidth ?

Amplifiers are designed to work over a band of frequencies outside of which the gain is attenuated. In some instances the requirement is uniform gain over a wide frequency range—as in high-quality audio amplifiers, certain measuring instruments, etc. A *wideband* amplifier is used in such cases and in general where complex signals, rich in harmonics, are to be handled.

Where, on the other hand, the requirement is to distinguish between a signal—possibly a very simple one—and unwanted noise, a more selective *narrowband* amplifier is appropriate.

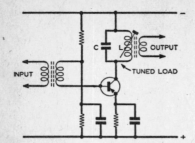

Fig. 46. An amplifier stage with a tuned circuit as the load to provide control over the bandwidth.

Bandwidth can be controlled by incorporating a resonant circuit or frequency-selective filter. Fig. 46 shows a tuned high-frequency amplifier in which a tuned circuit takes the place of a load resistor. The tuning capacitance C is selected to bring the circuit to resonance at the centre-frequency of the required band, with "fine tuning" by means of a variable magnetic core in the inductance L. As shown, input and output coupling is by means of mutual inductance.

Why is decoupling necessary ?

A number of amplifying stages in a piece of equipment normally have a common h.t. supply, across which the valves and components are connected. It is important to ensure that interaction, possibly causing oscillation and distortion, cannot occur between stages via the common supply. The appropriate decoupling is arranged by minimising the number of direct connections to the supply and by adding filtering components between each stage and the supply.

The aim is to restrict fluctuating currents to the stage in which they occur. The components (R and C) used for this in a typical valve stage, involving anode, screen grid and grid bias decoupling, are shown in Fig. 47.

What is meant by feedback, and how is it used ?

The term *feedback* is used to describe an arrangement whereby part of an amplifier's output is fed into its input.

58

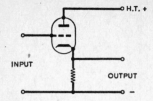

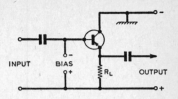

Fig. 50. The cathode follower
circuit.

Fig. 51. The emitter follower
circuit.

What is a cathode follower ?

This arrangement is best known for its very low output
impedance: it enables power to be fed into a low-impedance
load. The basic circuit is shown in Fig. 50. The input is
connected between grid and earth in the usual way, but the
output is taken from across the cathode resistor. No anode
resistor is necessary. A positive voltage at the grid increases
the anode current, which flows through the cathode load
to give a positive change at the cathode. The cathode thus
"follows" the grid in potential.

Not only is the output impedance low, but the input
impedance is very high and does not "load" the preceding
stage. Again, the cathode follower does not provide any
amplification (its gain is less than one). Finally the output
has the same phase as the input—the reverse of what
happens in a conventional amplifier stage. Note that by
adding a load resistor in the anode circuit two outputs are
available in opposite phase: such an arrangement is suitable
for feeding a push-pull output stage.

The emitter follower is the transistor counterpart of the
cathode follower. It is also known as the common-collector
or grounded-collector circuit. Again there is a high input
impedance and low output impedance, and the voltage gain
is less than one. A circuit is shown in Fig. 51.

61

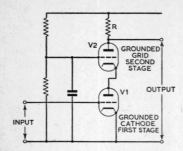

Fig. 52. Basic cascode amplifier circuit.

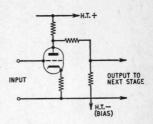

Fig. 53. Simple d.c. amplifier circuit.

What is a cascode circuit ?

The cascode circuit, shown in Fig. 52, consists of two triode valves—a stage with earthed (or "grounded") cathode directly coupled to a grounded grid stage, the two valves being in series across the h.t. supply. The arrangement provides the gain of a pentode valve while benefiting from the low noise of a triode. Consequently the arrangement has its uses as an input stage for amplifying very small signals. The same principle has been used for high frequency transistor amplifiers to reduce the effects of internal feedback in the transistors.

Explain the use of d.c. amplifiers.

The circuits so far described are designed to handle a.c. signals, and they incorporate coupling capacitors or transformers which block direct current while allowing a.c. to pass. It will be evident that a slowly varying signal will not pass any capacitor of practical size. Indeed even a large and expensive capacitor may block such a signal: it will store a big charge and will need time to dispose of it.

Therefore direct-coupled (d.c.) amplifiers are used to handle slowly varying or direct voltages. In these there is direct coupling between amplifying stages. As can be seen

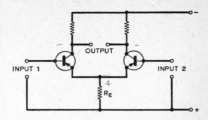

Fig. 54. The basic differential amplifier circuit, using transistors.

from Fig. 53, the h.t. of one stage is bound to appear at the grid of the next stage; therefore the grid must be biased with a large negative voltage to oppose this high positive voltage.

Unfortunately simple d.c. amplifiers are prone to *drift*; temperature changes alter the biasing and give rise in effect to an unwanted signal. It is essential to find ways of preventing th is. The obvious way is to use a compensating circuit. A widely used circuit that does this is the *differential* circuit—also known as a *long-tailed pair*—shown in Fig. 54. It consists of two transistors (as shown) or triodes with a common emitter or cathode connection. A steady current flows via R_E through the two transistors. With equal and in phase signals applied to the two inputs there is no output (change in current at the output terminals) since the current rises and falls equally in each. When, however, the inputs are unequal or out of phase, the current through one transistor increases as the other falls, and vice versa. The output is thus the difference between the inputs, and, since temperature changes will affect both transistors equally, temperature does not affect the output.

Transistors are in general use in d.c. amplifiers, and the use of alternate p-n-p and n-p-n silicon types helps to reduce drift and secure adequate gain. The silicon transistor causes less drift than the germanium type.

5

GENERATING SIGNALS AND PULSES

Describe the operation of a simple oscillator.

In Chapter 4 it was stated that positive feedback is the basis of oscillating circuits. If part of the amplified output of a valve is fed back to the input, in the correct phase, the result will be an increase in the gain of the valve, and the generation of continuous oscillations at a frequency determined by certain components in the circuit.

The output signal must be fed back in such a way that there will be phase inversion: then the feedback voltage will aid the input signal and the required amplification and oscillation will be built up. In practice it is not necessary to provide an input signal from a separate source, as the very small noise energy generated in the circuit will be enough to start the oscillation when the power supply is switched on.

Sine waves at a single frequency can be generated by a simple oscillator incorporating a tuned circuit. The resonant frequency of this circuit is the frequency of the oscillations.

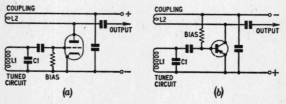

Fig. 55. (a) Valve LC oscillator circuit. (b) A transistor version, the tuned base LC oscillator.

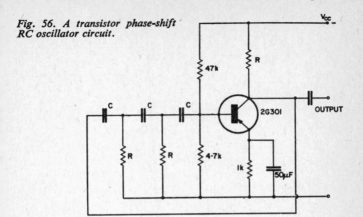

Fig. 56. A transistor phase-shift RC oscillator circuit.

One of a number of possible arrangements is seen in Fig. 55 (a).

Here, the tuned circuit consists of inductance and capacitance, and it will be seen that the inductor is in two parts—a coil in the anode circuit and, magnetically coupled to it, the grid coil. This constitutes the feedback path. Variations of current in L2 induce voltages in L1, and these are added to voltages already present, resulting in oscillation. Fig. 55 (b) shows a transistor version.

Describe a RC oscillator.

In this widely used type of oscillator the feedback is taken from anode to grid, or from collector to base in the circuit shown in Fig. 56, through a network of resistors and capacitors which determines the frequency of operation. In the circuit shown in Fig. 56 there are three *RC* networks, each introducing a certain phase shift, or alteration of phase, which depends on frequency. The circuit designer arranges for the networks to provide a phase shift of 180° (that is, phase reversal) at a particular frequency, and at this frequency oscillation occurs.

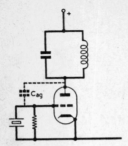

Fig. 57. An example of a crystal oscillator circuit, the Miller oscillator. The crystal is connected in the grid circuit, and positive feedback to sustain oscillation is provided by the anode to grid interelectrode capacitance (shown as C_{ag}) o the valve.

The frequency can be adjusted over a wide range if variable capacitors are used. A relatively elaborate arrangement based on the principle outlined here is incorporated in signal generators, which are used for test and measurement purposes. (There may be facilities for modulating the output by mixing it with a sine wave of lower frequency.)

What is a crystal oscillator ?

This type of oscillator makes use of the piezo-electric effect exhibited by certain crystal materials. As mentioned earlier, application of a voltage to such a crystal causes a slight mechanical deformation. If the crystal is held in a suitable way it will resonate mechanically, and its frequency stability is extremely good.

A crystal can be incorporated in an oscillator in a number of ways. For example, in the Miller oscillator (Fig. 57) the anode circuit is tuned to a frequency slightly above the crystal's resonant frequency: the crystal then oscillates at just below resonance point. This and other crystal oscillator circuits are used in a variety of equipments including scientific instruments, measuring equipment and radio transmitters. A frequency stability of one part in several million can be achieved: the best performance is obtained when the crystal (in practice a small wafer of the material) is fitted in a temperature-controlled housing.

66

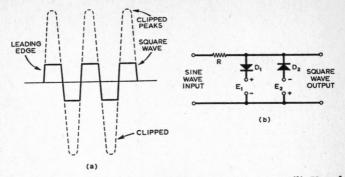

Fig. 58 (a) Clipping a sine wave to produce a square wave. (b) Use of diodes for clipping.

How is a square waveform produced ?

A sine wave can be converted into a square wave by removing the positive and negative peaks. It will be clear from Fig. 58 (a) that the peak value of the sine wave should be very much greater than the required square wave: this will ensure that the resulting waveform is as nearly square as possible (i.e. with a sufficiently steep leading edge).

A circuit for the removal of peaks is known as a clipper, or limiter. An example is shown in Fig. 58 (b). Diode D1 conducts when the input voltage rises above a predetermined clipping value—the bias voltage E_1. The resistor R is made larger than the resistance of the diodes when conductive. Therefore little change occurs in the output voltage until the input falls below E_1. At this point D1 no longer conducts, and the output voltage follows the input into the negative half-cycle. Then the other bias voltage E_2 is reached, and D2 limits the output voltage once more.

If a square wave of only one polarity is required, a series diode can be introduced. Pulses of one polarity will be passed, depending on which way round the diode is connected.

67

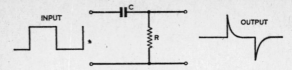

Fig. 59. Basic differentiating circuit.

What effect has the "time constant"?

We have seen that a capacitor cannot gain or lose its charge instantaneously if there is resistance in the circuit. The time constant is a measure of the rate at which a capacitor will charge through a resistor. It is numerically equal to RC (where C is in farads, R is in ohms and the time constant is in seconds). Again, time constant in the case of an inductive circuit is numerically equal to L/R (where L is in henrys). The time constant of the components in a waveform generating circuit determine the frequency of the waveform and affect its shape.

How is a spike waveform produced?

A *differentiating circuit*, having a restricted high or low frequency response, can be used. The values of resistance and capacitance (see Fig. 59) are chosen so that their time constant is much smaller than the time for one input pulse, and in this way the required distortion of the input waveform is obtained at the output.

If a square pulse is applied to the input, as shown, the capacitor first discharges (assuming it is already charged) and the full voltage is presented across the resistor. The capacitor then begins to charge again. If it is of small value and the resistance is also small, the charging will be rapid and the voltage across the resistor will fall rapidly. When the capacitor has fully charged the voltage across the resistor will be zero. When the rear edge of the pulse arrives, a similar process occurs but in opposite polarity. Thus the voltage of the capacitor is in effect a p.d. of reversed polarity across the resistor.

68

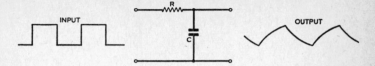

Fig. 60. Basic integrating circuit.

The discharge from the capacitor is of course relatively gradual, and the output shows a tailing off in voltage after the steep leading edge. For each square pulse there are positive-going and negative-going spikes, and in practice one of these would be clipped by a series diode circuit, leaving a succession of either positive or negative spikes.

How is a sawtooth waveform produced ?

An *integrating circuit* (Fig. 60) can be used for the production of a triangular or sawtooth wave from a square wave. The time constant is longer than in the previous example: it may be about equal to the time for a half-cycle of the input. However, if it is very long the waveform will be too smooth. In practical examples the capacitor is discharged rapidly via an electronic switch, e.g. a suitable oscillator. The most important application of this type of waveform is the provision of a timebase voltage for a cathode-ray tube (see later).

What other types of pulse circuit are there ?

Special circuits, involving switching between two operating states, are very widely used, for instance where high repetition frequencies and very short rise times are required. Transistors are now generally employed: two of these, with relatively simple coupling networks, are necessary.

The two operating states may both be unstable (the multivibrator); or one may be unstable and the other stable (monostable "flip-flop"); or both may be stable (bistable circuit).

69

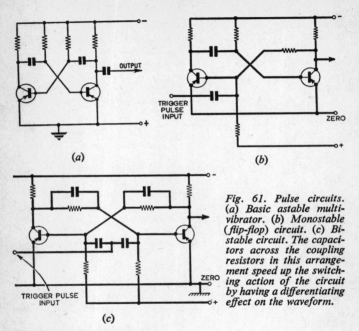

Fig. 61. Pulse circuits. (a) Basic astable multivibrator. (b) Monostable (flip-flop) circuit. (c) Bistable circuit. The capacitors across the coupling resistors in this arrangement speed up the switching action of the circuit by having a differentiating effect on the waveform.

Describe the multivibrator.

The multivibrator is an astable circuit: it has no stable condition. It was first devised (using valves) for the generation of square waves; and because a square wave is rich in harmonics the name multivibrator was applied to it. The basic circuit is shown in Fig. 61 (a). Feedback by cross-coupling is a feature of such pulse circuits. Here, positive feedback is provided by capacitative cross-coupling from the base of each transistor to the collector of the other. While one transistor is switching from being "off" to being "on", the other switches from "on" to "off".

The circuit is self-starting when switched on, as slight unbalance in the components causes the transistors to go into one of the unstable states. Each transistor is auto matically switched from a non-conducting state to a fully conductive (*bottomed*) state, and oscillation continues between the two unstable states—i.e. the oscillator is free-running.

A high p.r.f. is possible, but a limit is set by the switching times of the transistors. The multivibrator can be used in frequency multiplication: a particular harmonic can be selected and amplified so that a sinusoidal output is obtained.

Describe the monostable circuit.

The basic circuit is shown in Fig. 61 (b). It is stable in one of its two states. A "trigger" pulse flips the circuit into the unstable state, after which it flops back into the stable state (hence the often-used description "flip-flop" circuit). The astable circuit of the previous example can be made monostable by substituting a resistive cross-coupling for one of the capacitive couplings.

The circuit delivers one output pulse for each input pulse. It is suitable for pulse-shaping and pulse amplification. A square wave output can be obtained from a spike input.

Describe the bistable circuit.

The basic arrangement is shown in Fig. 61 (c). There are two stable states, and a trigger pulse has to be applied in order to make one transistor stop conducting and the other begin to conduct. The circuit remains in the second condition until a second pulse sends it back to the original condition. The astable circuit can be made bistable by substituting resistive cross-couplings for both the capacitive couplings (the capacitors shown in Fig. 61 (c) are not theoretically necessary but help the action of the circuit).

Since each trigger pulse sends the circuit into one stable state, there is one complete cycle of operation for every two input pulses. Therefore the p.r.f. of the output from either collector is half that of the input. It is a simple matter to convert the circuit to a binary counter.

71

6

RADIO AND TELEVISION

What is the principle of radio ?
The aim is to provide a "wireless" link for the transmission of information. This can be achieved by quite simple
means, and indeed in early experiments information was
conveyed in the form of electromagnetic waves without the
aid of the valves and other components that have since
become familiar.

It is necessary for radio-frequency energy to be generated,
fed to an aerial and radiated as electromagnetic waves into
space (the word "ether" was formerly used). As indicated
earlier, the waves do not require a material medium for
their propagation. In practice alternating currents are
generated at high frequencies, and the rapid changes in
current in the aerial set up electromagnetic fields which
become detached and move out into space.

That this is possible can be best appreciated by consider-
ing the effect of current reversal in a conductor, such as a
loop of wire. Passing a current through the loop (see Fig. 62)
sets up a magnetic field which becomes weaker the further
it reaches out from the conductor. A slow reversal of current
direction will produce a correspondingly slow building up
and collapse of this field.

On the other hand, a sufficiently rapid reversal will cause
a new field to be created before the previous one has had
time to die away. The new field, springing outwards from
the loop, will push with it what remains of the previous
one; some energy will be detached, moving out into space.

It is not difficult to visualise the result of continual high-

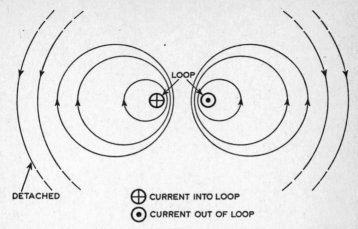

LOOP

DETACHED

⊕ CURRENT INTO LOOP

⊙ CURRENT OUT OF LOOP

Fig. 62. Radio waves will be radiated from a loop aerial by feeding high frequency a.c. to the loop: rapid reversal of the current in the loop results in the field around the loop being detached with each reversal in phase, as shown.

speed current reversals of the kind involved in r.f. energy. The radiation will be sustained, since a portion of the field will be detached during every cycle of operation. Again, the process can be thought of in terms of electrostatic fields which, as mentioned earlier, are created between spaced conductors.

Finally, if waves can be radiated into space in this way, they can also be received by an aerial of basically similar type. An e.m.f. will be induced in the aerial when an electromagnetic wave passes it.

Which modulation processes are used ?

The idea of modulation, and its role in sending information over distances, was introduced in Chapter 2. Both amplitude modulation (a.m.) and frequency modulation (f.m.) are used by broadcasting authorities and other professional and commercial users of radio.

73

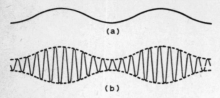

Fig. 63. Comparison of amplitude and frequency modulation. (a) Audio signal. (b) Carrier signal amplitude modulated by audio signal. (c) Carrier wave frequency modulated by audio signal.

In a.m. the amplitude of the high-frequency carrier wave (the r.f. energy) is varied in accordance with variations in the strength (amplitude) of the audio signals that are to be transmitted; while in f.m. the amplitude of the carrier is constant but its frequency is varied in accordance with the audio signal. These two systems are compared in Fig. 63.

Although a.m. is still widely used reception with it is in general affected by interference of various kinds. There are for instance interference from stations on neighbouring frequencies and noise generated by electrical equipment; and frequency response is affected by limiting the pass-band of the receiving circuits in order to make performance acceptable. F.m. broadcasting gives better standards of reception, though requiring slightly more elaborate equipment.

What are the merits of f.m. ?

Transmission at very high frequencies (v.h.f.: 30–300 MHz) can provide more freedom from most kinds of interference as well as an improved frequency response, even if the a.m. system is retained. However, only the combination of v.h.f. and f.m. leads to the required rejection of noise (e.g. man-made electrical interference) to enable interference-free reception to be possible.

A great deal of noise is similar in form to a.m. trans-

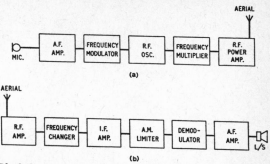

Fig. 64. Block diagrams of (a) an f.m. transmitter and (b) an f.m. receiver.

missions, so that by using f.m. broadcasting and incorporating a "limiter" in the f.m. receiver to prevent reproduction of a.m. signals, interference can be reduced. Apart from noise reduction and improved frequency response (to about 50–15,000 Hz under the best conditions), the user can benefit from better dynamic range—that is, the range from the quietest to the loudest sounds.

The section of the v.h.f. band occupied by f.m. transmissions in the U.K. is known as Band II, covering 87·5–100 MHz. A transmitter's range is rather limited at these frequencies and a relatively complex system of stations is needed to give national coverage.

The essentials of a transmitter and a receiver for f.m. are shown in simple form in Fig. 64.

How are radio signals demodulated ?

Demodulation (or detection, the traditional term) is one of the most important processes in reception. The purpose of the demodulator is to derive from the modulated carrier a copy of the modulation waveform, as free from distortion as possible. A single diode can be used in an a.m. receiver: it works in much the same way as the rectifier described in Chapter 3. A semiconductor diode may for example be

75

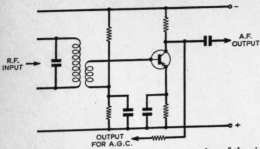

Fig. 65. Basic circuit of a transistor a.m. demodulator. If an RC combination with a suitable time constant is incorporated in the a.g.c. line, the a.g.c. potential consists of the average value of the signal at the demodulator, i.e. it does not vary with the a.f. information.

used with a potentiometer (the volume control) as its load.

A transistor may be used instead, in an arrangement such as that in Fig. 65. The emitter-base "diode" demodulates the signal, an amplified output being given at the collector. Power is provided as shown for automatic gain control—a common feature of modern receivers.

In the case of f.m. the demodulator is much more complex. One circuit, noted for its rejection of a.m., is the ratio detector. As shown in Fig. 66, there are two diodes and a double-tuned transformer with tapped secondary to which a third winding coupled back to the primary is connected.

Fig. 66. Basic ratio detector circuit (including final i.f. stage).

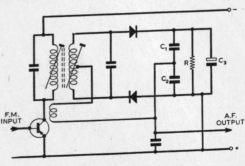

76

At the resonant frequency equal but opposite voltages appear at the opposite ends of the secondary and the third winding applies a reference voltage to the centre tap in series with these voltages; the diodes conduct equally and a steady current flows through R. Deviations in signal frequency cause phase shifts in the circuit which result in a signal of varying amplitude at a.f. appearing at the output. C3 provides stabilisation.

What is the principle of television ?

Television depends on the transmission of signals representing a complex pattern of light of varying intensities. Each part of the scene to be televised is treated as a separate light source, and the transmitted signal has to take account of the intensity of each. Successful reception depends also on persistence of vision: the information conveyed by the signal must be assembled on the picture tube screen with such speed that the viewer will see complete, moving pictures.

At the transmitting end the required electrical signals are obtained by *scanning* the scene from left to right in a series of lines. By scanning, the scene being televised is split up into a sequence of signals representing the light intensity of each part of the scene. These signals are reassembled in the receiver to form the picture. The speed of scanning is such as to produce a complete picture every $\frac{1}{25}$ second—fast enough for the eye and brain to gain an impression of continuous reception.

How does the camera work ?

The principle of operation of the image orthicon camera is illustrated in Fig. 67. (The word orthicon refers to the linear nature of its input-output characteristic.) An optical system focuses the scene on to the tube's photocathode, which emits electrons according to the light falling on it. The electrons travel to the target, where an electronic representation of the scene is formed. A low-velocity electron beam, emitted by an electron gun at the other end of the tube, scans the target.

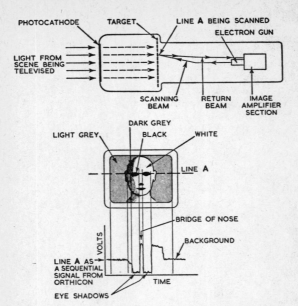

Fig. 67. Principle of operation of the image orthicon television camera tube.

Some of the beam electrons remain on the target, but many others return. The return beam, fluctuating in value according to the electrical value of the image on the target, is then increased in strength by an image amplifier before any further treatment is applied.

Scanning depends on both horizontal and vertical deflections of the electron beam, and this is achieved by applying sawtooth waveforms to coils fitted around the tube. This arrangement is similar in principle to that used in the receiver's picture tube. The basic scanning pattern is known as the *raster*.

The necessary waveforms are provided by timebase circuits: the line timebase deflects the beam horizontally and

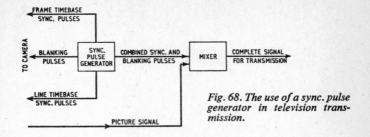

Fig. 68. The use of a sync. pulse generator in television transmission.

the frame timebase deflects it vertically. In current U.K. systems 405 or 625 lines are scanned per frame (complete picture) and frames are transmitted at the rate of 25 a second.

However, the process is modified by *interlacing* in order to reduce flicker. If half the lines are scanned in $\frac{1}{50}$ second and the alternate lines in the next $\frac{1}{50}$ second, the eye sees a frame every $\frac{1}{25}$ second. The picture quality is in this way improved without increasing the bandwidth occupied by the signal (i.e. any other method of improving quality would involve transmitting more detail, and this means a more complex signal needing more band space).

How is synchronisation achieved ?

There must obviously be synchronism between transmitter and receiver if the picture is to be reassembled correctly in the picture tube. In other words, there must be exact correspondence between the beam deflection in the camera and that in the receiver's picture tube. To ensure this it is necessary to transmit additional information, in the form of sync. (synchronisation) pulses, with the picture signal.

A sync. pulse generator is used (see Fig. 68) to produce a number of different pulses. At the end of each horizontal scan in the camera a line sync. pulse is generated and applied to the camera, and similarly at the end of each vertical scan

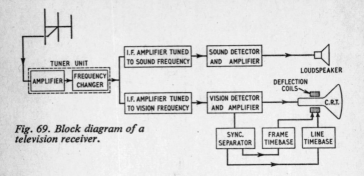

Fig. 69. Block diagram of a
television receiver.

there is a frame sync. pulse. There is also a series of blanking
pulses: these interrupt the picture signal while the electron
beam returns to the start of a line or frame, thereby sup-
pressing the flyback trace. The sync. and blanking pulses
are also combined and applied to a mixer, where they are
added to the picture signal to make the composite television
waveform which is transmitted.

The sound signal, obtained in the usual way with microphone and amplifiers, is not associated with the picture
signal. It is handled by a separate transmitter and received
in the same way as any other sound broadcast.

What are the main sections of a television receiver ?

Fig. 69 shows the principal sections of the receiver. First
comes a tuner unit, usually readily separable from the rest
of the receiver, and within this unit are the channel-
selection components (e.g. a turret tuner) and an amplifier
which handles both vision and sound signals. The frequency
changer stage changes the carrier frequency (of any channel)
to a fixed "intermediate frequency". This is a relatively low
frequency for which simpler and cheaper amplifying stages
can be devised than are needed to handle signals at the
transmitted frequency.

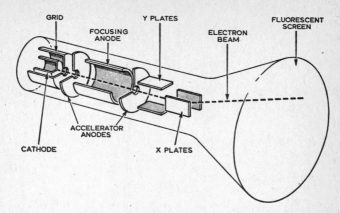

*Fig. 70. Principle parts of a cathode-ray tube with electrostatic deflection.
For television purposes electromagnetic deflection of the electron beam
is used, the X and Y plates being replaced by coils mounted externally,
to which the deflection currents are fed.*

The vision and sound signals are separated by tuned
circuits and demodulated; at this stage the carrier is removed and the signals are ready for final amplification.
The video output stage feeds the picture tube and the sync.
circuits. The sync. pulses are separated from the complex
waveform and applied to the timebase circuits which control
the scanning of the tube. Thus the timebase sections are
concerned with the movement of the beam, while the video
amplifier modulates the beam and determines the brightness
of the image on the screen.

What are the main features of a cathode-ray tube ?

The cathode-ray tube is widely used in one form or
another in radio and electronics work, particularly for test
and measurement purposes, as well as in television. It is
commonly used—in an oscilloscope—for examining the
waveform of alternating voltages. For the sake of clarity

the simplest kind of tube is shown in Fig. 70; a rather different type is used as a television picture tube.

A high-vacuum tube with electrostatic deflection and focusing is depicted. A gun system, comprising an indirectly-heated cathode, a control grid and focusing and accelerator anodes, projects an electron beam. This beam passes between two pairs of deflector plates, the X and Y plates, and by applying suitable voltage waveforms to the plates the beam can be directed as required to any position on the screen.

The fluorescent screen is made by coating the inside face of the tube with a compound which emits light when struck by the electron beam. Brightness is controlled by varying the negative voltage on the grid (or cathode), and focusing is effected by adjusting the voltage on one of the anodes. Vertical deflection is obtained by applying suitable voltages to the Y plates, and horizontal deflection is similarly achieved by applying voltages to the X plates.

In contrast, television picture tubes depend on magnetic (instead of electrostatic) fields for deflection. The coils which produce these fields are mounted externally around the tube leaving only the gun and focusing system inside the tube. Line and frame deflector coils, the counterparts of the X and Y plates, are split into halves and mounted each side of the tube neck. Focusing may alternatively be provided by a permanent magnet. These arrangements permit beam deflection over a very wide angle, and consequently a short tube is possible.

Tubes for colour television are more complex as they have three electron guns, one for each of the three colours (red, green and blue) and more involved deflection arrangements. These are fully described in *Questions and Answers on Colour Television*.

7

CONTROL ENGINEERING

What are the main features of control systems ?

In widely used control arrangements a transducer of some kind produces signals which are related to the quantity to be controlled—the speed of an electric motor, for example. The signals will probably require amplification; after this they are used to carry out some control operation.

The use of a tachometer for checking speed may be instanced: this device has to be read by an operator, who subsequently (after inevitable delay) takes appropriate action if he finds that this is necessary. A tachogenerator, on the other hand, is coupled to the motor in the same way but gives an electrical output related to speed. This output can be used to initiate automatic corrective action.

Although amplification of signals is often necessary, the problems of handling complex waveforms (as in radio) do not arise in normal control systems, except where pulses are being generated. Slowly varying signals have to be dealt with, and d.c. amplifiers are used. Special precautions are taken against drift, and low-frequency oscillation, known as hunting, is minimised by error-correcting circuits.

Thyratrons and silicon controlled rectifiers (s.c.r.s) are in general use as switching devices in control circuits for d.c. motors, d.c. power regulating systems and alternators. Compared with the thyratron, which it is gradually replacing, the s.c.r. has a very low forward voltage drop in the conducting state and is highly efficient. It is very quick-acting, operates in high temperatures and, of course, has no heater.

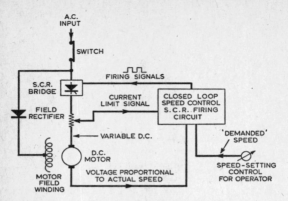

Fig. 71. Control system for a small motor. As we saw in Chapter 3, the s.c.r. bridge rectifier circuit will control the d.c. to the motor in accordance with the timing of the triggering, or firing, pulses.

Give an example of motor control.

Variable-speed drives are required in many industries and for a variety of motor-driven machines. Flexible yet reasonably simple control arrangements are made possible by thyratrons or s.c.r.s (also called thyristors). Such devices can be fired, i.e. made to conduct, at any point during the positive half-cycle of an alternating current which the device is being used to rectify, so enabling a continuously variable d.c. output to be obtained from a constant a.c. input.

A control system for a small d.c. motor (say 4–7 kW, about 5–10 h.p.) is shown in Fig. 71. The motor field is excited at constant voltage by rectified a.c. The speed of the motor is altered by adjusting its armature voltage by means of a s.c.r. bridge circuit which is switched from non-conducting to conducting states by low-power firing pulses.

The timing of these pulses controls the instant during the a.c. cycle when the s.c.r.s start to conduct and hence controls the average output voltage level. A simple control circuit adjusts the timing of the pulses according to the requirements of the operator, who uses a speed-setting potentio-meter. Somewhat similar systems are used for motors up to 150 kW (about 200 h.p.).

The Ward–Leonard system is used for fine speed control of motors of much higher power. In this relatively elaborate arrangement the armature of the main motor is supplied by the generator of an auxiliary motor-generator set. The latter always runs in one direction, and reversal and control of the main motor is arranged by suitable switching and resistors in the generator field circuit.

The finer measure of control offered by electronics is introduced by supplying the generator field from thyratrons or s.c.r.s. A special facility is provided by this arrangement: regenerative braking (that is, braking by purely electro-magnetic means) is applied when the set is adjusted to a lower speed. Firing circuits for s.c.r.s may give a rectangular pulse, a spike or a series of pulses, e.g. from a multivibrator.

How is automatic control achieved ?

In a conventional attended installation the operator would observe errors in motor speed (or some effect due to the errors) and then take action to correct them. The first step towards automatic control is to obtain feedback of information about errors. This information is then used to apply corrective action. Thus a control "loop", formerly completed by the human operator, is formed by electronic means and automatic control becomes possible.

If, as suggested earlier, the motor drives a transducer, which provides a signal related to speed, this signal can be compared with a reference level. An error signal—the difference between the signal the transducer provides and the reference level—is thus obtained, and this can be used for control purposes.

In the closed-loop system shown in Fig. 72 the motor speed is monitored by a tachogenerator the output of which, Vt, is proportional to speed. This voltage is compared with the reference voltage, Vr. The deviation, d, between Vt and Vr is measured and amplified by the deviation amplifier.

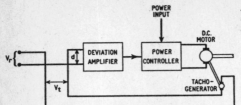

Fig. 72. A closed-loop feedback automatic control system.

The signal thus obtained is used to control the power controller, for example an s.c.r. rectifier arrangement as in the previous examples, which determines the input to the motor.

Can voltage control be applied to alternators ?

A very stable voltage may be of great importance in, say, the industrial plant which an alternator is supplying. This stability may be required despite fluctuating loads or even sudden peaks in the demand on the system. Here also it is possible to use a form of closed-loop arrangement.

The main requirements are rectification of the a.c. supply and comparison of the d.c. output with a stable reference source, such as a gas-filled tube. The error signal is the difference between the supply voltage and the reference source; after amplification, the signal is fed to the grids of several parallelled valves connected in the machine being controlled's field circuit and used as control devices. The number of valves required depends on the size of the field current.

Can alternators be synchronised by electronics ?

Automatic synchronising is increasingly used in industrial generating installations. Special arrangements made to meet peak loads provide excellent examples of the way in which electronic techniques become allied to modern electrical engineering practice. For instance, factories having their own diesel-driven generating plant can arrange for automatic start-up and synchronisation to the mains during periods of heavy demand. Reducing peak demand in this way avoids penalties imposed by the electricity supply authorities.

A practical synchronising unit senses voltage differences between running and incoming supplies (or between a machine and the mains) and delivers correction signals. When the difference comes within small limits the unit operates relays which in turn close the circuit-breaker. Accurate phase sensing is necessary, since serious damage can be caused by inaccuracies in this respect. A comparator measures the relative phase rotation between running and incoming supplies and controls the governor on the diesel engine to bring the speed to the right point for synchronisation.

The electronic elements of such a system, built entirely around semiconductor devices such as silicon transistors and diodes, include special kinds of amplifier and pulse-generating circuits. The advantages are of the kind found in many automatic systems—quick operation and the virtual elimination of mismatches which may arise through errors in manual control.

How can a series of different actions be controlled by a closed-loop system ?

By programming the reference source, i.e. replacing the "stable" reference source by a reference source consisting of a series of signals, the operation being controlled is given a series of orders so that it performs a series of operations automatically. A simple way of providing a series of orders for this purpose is to record the appropriate signals on magnetic tape. In this way machine tools can, for example, be automatically controlled.

8

COMPUTERS

How many types of computer are used ?

There are two main types—digital and analogue—and in general they are used to make calculations and solve problems which are difficult or tedious to deal with by manual methods.

For instance, digital computers are used in commerce and industry for extensive arithmetical calculations of a routine kind which would otherwise require very considerable clerical effort. *Data processing* is the term normally used to refer to such work. Analogue computers are used as aids to research and design, for example in speeding up complex laboratory investigations. Moreover the availability of computers means that some problems, especially in scientific research, can be tackled for the first time.

The outstanding difference between the two main types can conveniently be explained as follows. The digital computer can undertake any problem that can be expressed as a series of arithmetical operations. Such problems do of course embrace an almost unlimited number of tasks which do not appear to take even a remotely "mathematical" form. This type of computer works on numbers, and its operation rests firmly on the fact that information fed to it can be converted into a simple system of numbers which are represented in the computer by electrical pulses.

In contrast the analogue computer is used for work on quantities such as electrical voltages or mechanical movements whose magnitudes may fluctuate continuously with time, and it deals with mathematical equations.

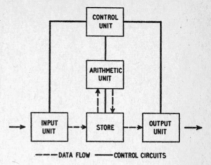

Fig. 73. Basic parts of a data-processing system.

The word "analogy" means correspondence in certain respects between things that are otherwise different, and this in fact is the key to analogue computer operation. It suggests that problems can be simulated, and indeed the computer is used in this way—as in the design of large structures and industrial plant. The effect of varying a number of design factors can be studied: the user is able to devise a "model" of the structure or plant under investigation.

How does the digital computer operate ?

The digital computer can do arithmetic at very high speeds and can make logical decisions when faced with alternative courses of action. It does not require constant supervision while such tasks are performed. However, these benefits become practical reality for the user only if the computer is correctly instructed: the work to be done by the computer must be converted into a *programme*, prepared in such a way that the data can be processed in simple step-by-step fashion. One requirement is the conversion of decimal digits into binary code.

Fig. 73 shows the flow of data in a computer used for commercial data processing. The main parts are: an input unit which enables both the instructions and the data to be fed into the computer; an output unit, to communicate the results of the calculations; a storage unit, or "memory", which retains data and instructions during the computing process; an arithmetic unit, which performs the calculations; and a control unit which exercises overall control in accordance with the programme.

The word "hardware" (an Americanism) is often used to refer to the various printing machines, tape and card punches and other equipment used with a computer. Indeed, the electronic assemblies at the heart of the computer can equally well be called hardware. Materials concerned with the programming of the computer are known as "software". Auxiliary units (essentially hardware) associated with a computer may also be referred to as peripheral equipment.

What is the binary code?

The decimal notation, using the ten figures 0 to 9, is involved in most calculations, including those which computers are required to undertake. Although a machine can be made based on the use of these ten figures, a simpler system is needed for general use in electronic computing.

This is because of the nature of the electronic devices used. Reference has already been made to the switching function of semiconductor devices and other components: most of them are two-state devices which are either "on" or "off". Therefore the binary notation is used in computing, since it has only two figures, 0 and 1, which can be represented by these "on" and "off" states. Although it may appear that calculations with such a code must be ponderous, the arithmetic circuits in a computer can change their state extremely quickly—many millions of times a second.

The precise form of coding employed varies a good deal, but the principle involved in obtaining the binary equivalent of a decimal number can easily be illustrated. The decimal number is repeatedly divided by 2, and the remainder at each stage is indicated by 0 or 1. Therefore the binary equivalent of 55 is obtained as follows:

55	27	13	6	3	1	0
	1	1	1	0	1	1

The binary number is given by the remainder column from *right to left* and is therefore 110111.

In an alternative system, individual decimal digits are converted into a straight binary code; four binary places are used for each digit, giving equivalents as follows:

9	8	7	6	5	4	3	2	1	0
1001	1000	0111	0110	0101	0100	0011	0010	0001	0000

In this form of coding the number 55, read directly from the equivalents, is 01010101.

Mention of *bits* of data will be seen in descriptions of computers. The word is derived from *bi*nary dig*its* and thus refers to the binary code discussed above.

How is information stored ?

Storage devices have developed very considerably since digital computers were first introduced. An early type of store was a cathode-ray tube in which the screen coating was split up into small charge areas. In more recent times magnetic discs and drum stores have been used.

In the drum type a rotatable metal cylinder is coated with a ferrous oxide and held in suitable bearings. Recording and reading heads are placed near its surface. Information in pulse form, passed through the head winding, induces a magnetic flux and hence records information on the drum in the form of very small magnetised areas. Reversing this process enables data to be extracted, after which they are amplified.

A drum less than a foot long may store 300,000 binary digits, and the time of access to this information may be only a few milliseconds. The principle involved here will be familiar to users of tape recorders. Indeed, magnetic tape stores are often used to provide large capacity. The tape is wider than the 6·35 mm ($\frac{1}{4}$-in.) variety used on domestic recorders and numerous tracks are recorded on it. Apart from this, it will be evident that storage capacity is extended by using one tape after another on the one machine.

The magnetic core store is another important technique now much used. A large number of small ring-shaped ferrite cores are threaded by wires along which pulses travel, and changes in magnetic state represent the "0" and "1" of the binary notation. The advantage of such an arrangement —apart from absence of moving parts—is the short access time, measured in microseconds.

How is arithmetic performed ?

The arithmetic unit in a digital computer performs addition, subtraction and multiplication of the coded data. The "logical elements" which undertake this work are basically very simple switching and amplifying circuits since, in general, they have only two operating conditions. However, a very large number of such circuits have to be interconnected, and they are extremely fast in operation. For these and other tasks in a computer 15,000 semiconductor devices may be used.

How do logic circuits work ?

Circuits known as AND and OR gates provide the simplest examples of computer logic and are also used in other kinds of electronic equipment. A gate is a switch in this context.

Circuits using diodes are shown in Fig. 74. Assuming positive input pulses, the AND gate gives an output only when pulses are applied simultaneously to each input; but the OR gate gives an output if a pulse is applied to either

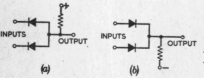

Fig. 74. Simple two-input diode gate circuits. Assuming positive-going input pulses, (a) gives the AND function and (b) the OR function.

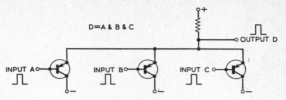

Fig. 75. Transistor AND gate.

one or both of the inputs. However, if the input is made negative, the circuits reverse their functions.

A transistor AND gate is shown in Fig. 75. Other functions are also needed: for instance a NOT element gives an output only if no input is applied. As an example of the mathematical relevance of these circuits, it will be seen that the AND gate in effect adds, since if a load resistor is connected across its output the voltage across this will be the sum of the voltage signals applied to it. Actual arithmetic arrangements, however, are a lot more complicated than this, and are based on binary notation. Logic circuits are generally employed to gate the outputs of groups of bistable circuits (see Fig. 61 (c)). These each divide by two, as we have seen, but can be arranged in groups to register changes in binary notation inputs.

How does the analogue computer work ?

Circuits are devised for the direct execution of mathematical operations, typical processes being summation, integration and differentiation. The input, possibly derived from measuring transducers or instruments, is applied to a voltage amplifier incorporating negative feedback networks. Different basic mathematical processes are performed by altering the feedback elements. Special amplifier requirements include high gain and good stability, so that widely differing feedback networks can be introduced without provoking instability. The differential amplifier discussed in Chapter 4 is much used in this type of work. Indeed it is often called an *operational amplifier* as it will perform mathematical operations.

Analogue computers are being used for such work as investigating stresses in aircraft, ships and large engineering structures, and for setting up models of complex industrial installations in order to study the effect of changing certain design or operational factors. They are also used for training and educational purposes.

Are digital computers also used in research ?

The digital computer can be used for a variety of investigations, particularly where, as is so often the case, a great deal of time-consuming calculation or analysis of experimental results is necessary. An interesting feature is the use made by scientists of knowledge gained by others working in quite different fields. The outstanding example is the non-mathematician's use of mathematical processes which he does not understand: a suitably programmed computer enables him to employ them.

Uses for computers range from fundamental research in physics and chemistry to problems in technology, where it may be required to try out the effects of designing circuits in many different ways or to analyse the results of tests on the reliability and life of components.

9

SOME OTHER ELECTRONIC TECHNIQUES

What is radar ?

Radar is essentially an electronic system of navigation and direction finding, using a beam of electromagnetic waves. Its name is derived from *ra*dio *d*irection *a*nd *r*ange-finding. The relevant techniques stem from studies of the behaviour of radio waves undertaken about forty years ago. By the start of the Second World War these techniques had been considerably developed by British scientists for defence purposes. Subsequently great advances were made with radar as a navigational aid as well as for defence. Radar has an especially important place in the history of electronics since many of the techniques originally developed by radar engineers to meet urgent military needs subsequently made possible wide advances in the use and application of electronics.

Suitably projected energy, in the form of extremely short but powerful pulses, is reflected by any object which it may encounter (an aircraft, say, or a ship) and returns along the original transmission path. The energy is received and displayed by electronic means so that information about the presence, distance and movement of objects is obtained.

The direction of an object is given by the direction of the beam. The time taken for a pulse to travel to the object and back is measured; and, since the velocity of electromagnetic waves is known, the distance of the object can be calculated.

In developing radar, major aims have been increased

accuracy and sensitivity. Hence there has been a need for equipment, particularly new types of magnetron valve, to provide pulses of greater power, and it has been necessary to reduce considerably the wavelengths of operation. The transmission of *microwave* energy eventually became possible: the wavelengths are measured in millimetres (see Table 2 in Chapter 2). Low noise systems have also been a major aim, since only little of the energy originally transmitted returns, and that which does may not stand out amongst the background noise always present.

How is microwave energy produced ?

The radar pulses are very short bursts of oscillation, produced at the rate of about a thousand a second. Often the source of these is a cavity magnetron, a special type of valve which enables high peak powers of a megawatt or more to be generated at frequencies in the s.h.f. region. The magnetron was born of experiments to determine the influence of strong external magnetic fields on the behaviour of a certain kind of diode.

The operation of the magnetron depends on the behaviour of electron clouds in a combination of electric and magnetic fields. In radar the magnetron may be used in conjunction with a thyratron which, as mentioned earlier, is in effect a quick-acting control device—a switch, in fact. A d.c. field in the magnetron is provided by the thyratron during its conduction period; the magnetic field is supplied by a large permanent magnet. The arrangement is shown in Fig. 76.

The movements of the electrons given off by the magnetron's cathode and influenced by both the magnetic and electric fields are somewhat complex. The fields are arranged in such a way that the electrons travel round in cycloidal paths in the space between anode and cathode. Resonant cavities around the anode are set into oscillation by this action (each cavity behaving as a tuned circuit), and the rotating movement of the outer electrons maintains this oscillation.

During the interaction between the electron streams and

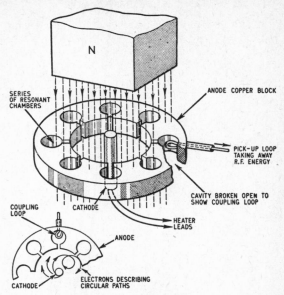

Fig. 76. Principle of operation of the magnetron microwave oscillator.

the anode, some of the electrons lose their velocity at the cavities and are collected by the anode, but others gain velocity and return to the cathode. This source of energy can be tapped at one of the cavities: a wire coupling loop is shown in the illustration. The power is then conveyed by waveguide to the aerial system.

The thyratron also requires a "firing" pulse to make it conduct. In practice a voltage spike is applied to its grid, and a suitable waveform can be provided by a multi-vibrator, a type of pulse generator which was mentioned earlier.

Other methods of generating microwave energy are also used. The high-power klystron, another type of microwave valve, is in use. The preferred system today is to use micro-wave amplifiers to amplify suitable waveforms, building up the signal waveform from a low level to an appropriate power for feeding to the aerial.

Describe the waveguide and aerial.

In its commonly used form a waveguide is essentially a pipe of rectangular section along which electromagnetic energy can travel. At the centimetre wavelengths used for radar this arrangement is more efficient than the other, more familiar methods—i.e. suitably constructed cables—used elsewhere in electronics.

The conditions under which the energy can travel in the waveguide are illustrated in Fig. 77: the electric field is perpendicular to the walls and the magnetic field is tan-gential to the walls. The waveguide can be constructed so that it turns corners in following the route from transmitter to aerial (see Fig. 78).

It is required to transmit the energy in a beam, and to

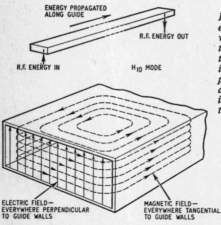

Fig. 77. Magnetic and electric fields inside a waveguide. The energy to be transmitted along the guide is coupled to it by means of a small probe or loop, which acts as a miniature aerial, and is extracted in the same way.

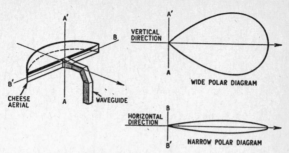

Fig. 78. A typical "cheese" aerial for radar transmission and reception, showing the polar characteristics of this form of aerial construction.

do this a special type of aerial is needed. One example is shown in Fig. 78. The use of such short wavelengths means that the aerial can be made fairly small.

The ability of the aerial to concentrate its output in a beam is especially important, and in this connection the *polar response* is of interest. (This property, i.e. the directional characteristics, is important throughout electronics, wherever there are devices—microphones or loudspeakers, for instance—which project or receive energy.) The polar response is made narrower by widening the aerial dimension in a particular plane; the response for one possible arrangement is shown in Fig. 78.

How is radar information displayed ?

Information about a detected object within the range of the radar is displayed on a cathode-ray tube. One widely-used arrangement is known as *plan-position indication*, in which a large-screen tube gives a direct indication of the origin of echoes—the reflected pulses—received by the radar.

As shown in Fig. 79, the scan is radial: the trace on the screen starts at the centre, where a spot indicates the transmitter's position, and rotates in circles. The aerial is rotated by an electric motor, and the trace on the screen is made

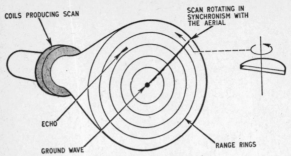

Fig. 79. *Typical p.p.i. display.*

to follow this by synchronising the rotation of the tube's scanning coils with the aerial.

The tube has "high persistence": a brightening of the trace, when an object is detected, persists for a time after the disappearance of the signal which caused it. In the absence of an echo signal, only the transmitter position is indicated; but on arrival of a signal and its application to the grid of the tube, the trace is suddenly brightened.

The bright spot is "painted" on the screen at a radius which corresponds to the range of the detected object. By its nature the display also provides information about the bearing of the object.

How is ultrasonic energy generated and used ?

Ultrasonic energy, at frequencies above about 20 kHz, is generated by two basic kinds of transducer. The piezo-electric transducer, based upon a property possessed by certain crystals, has already been mentioned. The other type is the magneto-striction transducer. Magneto-strictive effects caused by the application of a magnetic field to a ferromagnetic material result in a change in dimensions of the material.

The transducer is coupled to the medium into which it is to direct its energy. In marine applications it is necessary to focus the beam from one transducer and to use a second transducer to receive reflections. In this way the radiations can be used for depth-sounding, underwater navigation and the detection of fish shoals.

Ultrasonic equipments, based on focused beams of ultrasonic energy, are widely used for the non-destructive testing of materials, the inspection of castings and metal strip for flaws, the checking of thickness and other industrial purposes. The reflected ultrasonic signals may be displayed on an oscilloscope or recorder. Welding, drilling, soldering and other operations can be done by ultrasonics.

The cleaning of intricate and delicate components and mechanisms is a task which ultrasonic energy undertakes very efficiently. The cleaning action is largely due to *cavitation*, an effect due to the propagation of high-power ultrasonic waves in a liquid (such as a cleaning solvent or water). The molecules of the liquid are accelerated and extremely small vacuum bubbles continually develop and collapse, exerting a scrubbing action on solids which they encounter.

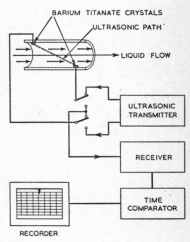

Fig. 80. An ultrasonic flow-meter. Ultrasonic energy is alternately transmitted and received by each piezo-electric transducer. The time taken for the energy to pass from one transducer to the other in both directions when compared indicates the rate of liquid flow.

101

An interesting industrial use is the measurement of liquid flow. There are several possible ways of basing a flowmeter on ultrasonic principles, and one of these is illustrated in Fig. 80. One piezo-electric transducer is used as a transmitter and the other as a receiver, and the time taken for a pulse to travel between them is measured. A pulse directed against the direction of flow takes longer to reach the other transducer than a pulse sent in the other direction; therefore the difference between these times is proportional to rate of flow of the liquid.

How important are electronic techniques in medicine ?

Electronic techniques are well established in medical research. Examples range from biochemical analysis, using photoelectric and semiconductor devices, to cancer research using electron spin resonance equipment. Apart from this, electronic methods, involving instrumentation of an advanced and specialised kind, are making important contributions to diagnosis and treatment in hospitals. X-ray examination and the use of closed-circuit television are of course familiar.

The use of electronics for diagnosis is on the whole a recent development, although a few methods are long established. The body generates very small electrical impulses which, suitably amplified and interpreted, provide information about the condition and behaviour of muscles, nerves and organs. Important aids include the electrocardiograph for examination of the heart's behaviour, and the electroencephalograph for the study of the brain.

Such equipments detect the very small voltages (a few microvolts, perhaps), known as bio-electric potentials, which are set up by muscular and other actions in the body. Electrodes are applied to the body and the signals they detect are amplified and then recorded or displayed as appropriate. Other instruments, such as the electronic stethoscope, detect and amplify vibrations in the audio-frequency range.

The development of semiconductor devices and, with

them, the miniaturisation of equipment is of general interest in medicine. For example, measuring instruments are reduced in size so that they can easily be carried on the person for monitoring purposes. Again, instruments of certain kinds are much more useful if they are battery-powered and easily portable.

Electronic equipment is now used for such very different purposes as continuous monitoring of blood pressure and the stimulation of reactions in the body. The latter application is one of the most important of recent times. A transistorised receiver can for example be inserted in the heart muscle of a patient whose heart-beats are irregular; this stimulator is controlled by a transmitter worn by the patient in his clothes. The device may be in operation for years without attention—evidence of the standard of reliability that can now be achieved.

Are radio techniques used in medicine ?

One of the smallest devices used in diagnostic medicine is the radio pill. This is essentially a radio transmitter and contains, in a housing less than an inch long, a transistor oscillator, battery, aerial and other parts. The pill is swallowed and transmits data from the stomach and intestines concerning acidity, temperature, pressure and other factors; and these can be assessed by the physician and related to other observations of the patient's condition. Frequency modulation is employed for the transmission of data. Temperature changes, for example, can easily be transmitted as a variation in frequency of the signals; and acidity in the stomach can be detected by suitable electrodes on the pill's surface, the pick-up voltages being applied to the oscillator to modulate its frequency. A receiver is placed near the patient, and the signals are recorded for later examination.

What is micro-miniaturisation ?

Electronic circuits and equipment are steadily becoming smaller, and many of the most recent innovations stem from efforts to meet the requirements of the armed services and space researchers. Some of the techniques are put to effective use in computers, where size is to a great extent related to operational efficiency; others are used by designers of industrial control systems.

In general the aim is to produce very small and reliable *integrated circuits* in which a variety of components, such as transistors, resistors and capacitors are formed together with their interconnections as film-like layers. The techniques generally stem from transistor technology. A number of thin-film circuits, each representing an amplifier, oscillator or some other functional stage, can be built up, interconnected and encapsulated in a small housing. In some equipments air conditioning, or a more elaborate system of temperature control known as *environment stabilisation*, is used in order to increase reliability.

Thin-film circuits may be formed on a ceramic, glass or alumina base—the substrate—only a few thousandths of an inch thick. A vacuum-deposited film can be treated to represent resistance, capacitance and so on at appropriate parts of the surface. The elimination of hand-made joints and connections within a functional stage or circuit makes a big contribution to reliable service.

How small are these circuits ?

One example is an amplifier formed in an area of about a square inch. A complete item of equipment may consist of a number of such circuits and, when housed, might have a volume only a fiftieth of that occupied by the more familiar transistorised equipments laid out on printed wiring boards. It has become possible to encapsulate 50,000 components in a volume of one cubic foot, and the techniques for much higher component densities are already available.

Thin-film circuits are likely to be employed in due course in domestic radio and electronic equipment. It is largely a question of waiting until such a time that they can be produced more cheaply than more conventional techniques. At present they contribute to light weight and reliability in equipment of the most advanced kind, such as experimental automatic take-off and landing systems for aircraft.

However, quite different and even more advanced methods are being investigated. The outcome may be the quantity production of truly "solid-state" equipments. This involves the formation of functional elements within crystals of semiconductor material, so that a further reduction of size becomes possible. Traditional ideas of circuits and current flow have little relevance here, since operation depends on phenomena within domains of molecules.

Describe some recent developments.

Amongst recent developments are the parametric amplifier, the maser and the laser. The first two are of importance as low-noise devices, useful in such applications as radio-astronomy and space communications where minute signals are involved. The laser is a source of coherent light, which is capable of modulation providing new possibilities in the field of communications.

In the parametric amplifier an oscillator known as the pump provides an output which varies one of the elements —e.g. a variable capacitance diode—in the amplifier circuit in such a way that the pump output energy is transferred to the signal being amplified. The maser (*M*icrowave *A*mplification by *S*timulated *E*mission of *R*adiation) is a solid-state device working on quantum principles. A crystal of suitable material, e.g. ruby, absorbs energy from an h.f. oscillator and is stimulated by the input signal to emit this energy at the signal frequency. The device requires a very strong magnetic field and works at temperatures several hundred degrees below zero.

The laser (*L*ight *A*mplification by *S*timulated *E*mission of *R*adiation) operates on similar principles to the maser, but is of importance in that it produces a wavetrain of nearly pure monochromatic light which can be focused into a narrow beam and modulated to provide low-power communication over vast distances.

The laser has been used in eye surgery and can also be employed, at relatively high powers, to drill or otherwise machine metals. Lasers are available using gas or semi-conductor material as the emitting medium.

What is the "field effect" transistor?

The field effect transistor differs from other types of transistor mainly in that control over the flow of current through it is achieved by applying a voltage to one of its electrodes. In this respect its operation is more akin to that of a thermionic valve than to that of other types of transistor. There are two main types of field effect transistor. The junction type consists of a bar of semiconductor (p or n) material with regions of the opposite type semiconductor (n or p) at the sides. Connections, called the source and drain, are made to each end of the bar, while the side regions are connected together and a connection made to them called the gate. There is thus in effect a single pn junction. The supply voltage is connected across the source and drain so that a current flows through the bar. The control voltage is applied to the gate to reverse bias the pn junction. The field established in the bar by the control voltage controls the current flowing through the bar. Thus the current flow varies in accordance with the voltage applied to the gate, and an amplified output will appear across a load resistor connected in the drain circuit. The device has a high input impedance, in this respect again being like a thermionic valve, and low noise factor. With early field effect transistors manufacturing problems led to difficulties in producing controlled characteristics, and this discouraged its use: these problems have however been resolved by improved manufacturing techniques.

The other main type is the insulated gate field effect transistor, also known as the metal oxide semiconductor transistor. With this type source and drain connections are made to separate regions of the same type of semiconductor material diffused into a block of the opposite type of semi-conductor material; the surface is covered with an insulating layer of silicon oxide, and a gate connection deposited on top of this. With the supply voltage connected between the source and drain, little current flows as there are two pn junctions back-to-back. Biasing the gate increases the conductivity between the source and drain regions, thus controlling the current through the device.

Field effect transistors have good electrical characteristics and are in many respects easier devices for circuit application than other types of transistor. They are thus likely to be of increasing importance. Their first application was as low-noise input stages for signals in the microvolt region, but a number of designs for v.h.f. and u.h.f. radio tuner units using them have recently been announced. Power types have not so far been developed.

USEFUL UNITS AND SI CONVERSION FACTORS

SI Base Units

Length	metre	m
Mass	kilogram	kg
Time	second	s
Electric current	ampere	A
Temperature	kelvin	K
Luminous intensity	candela	cd
Substance	mole	mol

Multiples and Sub-multiples

M = mega ($\times 10^6$)
k = kilo ($\times 10^3$)
c = centi ($\times 10^{-2}$)
m = milli ($\times 10^{-3}$)
μ = micro ($\times 10^{-6}$)
n = nano ($\times 10^{-9}$)
p = pico ($\times 10^{-12}$)

Supplementary Units

| Plane angle | radian | rad |
| Solid angle | steradian | sr |

Imperial to SI Conversions

Length

1 in = 25·4 mm
1 ft = 304·8 mm
1 yd = 0·9144 m

Area

1 in^2 = 645·16 mm^2
1 ft^2 = 0·092 903 m^2
1 yd^2 = 0·836 13 m^2

Volume

1 in^3 = 16 387·1 mm^3
1 ft^3 = 0·028 316 8 m^3

Velocity

1 in/s = 25·4 mm/s
1 ft/s = 0·304 8 m/s
1 mile/h = 0·477 04 m/s
= 1·609 344 km/h

Mass

1 lb = 0·453 592 37 kg
1 ton = 1016·05 kg

Density

1 lb/in^3 = 2·767 99 $\times 10^4$ kg/m^3
1 lb/ft^3 = 16·0185 kg/m^3

Force

1 lbf = 4·448 22 N
1 dyne = 10^{-5} N

Pressure

1 lb/in^2 = 6·894 76 kPa

Power

1 hp = 745·70 W

Temperature

$°F = (9/5°C) + 32$
$K = °C - 273$

108

Derived Units

Frequency	Hertz	Hz	cycles/second
Force	Newton	N	kgm/s²
Work, energy, heat	Joule	J	J = N m
Quantity of electricity	Coulomb	C	C = A/s
Potential difference, e.m.f.	Volt	V	V = W/A
Capacitance	Farad	F	F = A/s/V
Magnetic flux	Weber	Wb	Wb = V/s
Magnetic flux density	Tesla	T	T = Wb/m²
Inductance	Henry	H	H = V s/A
Luminous flux	Lumen	Lm	Lm = cd sr
Illuminance	Lux	Lx	Lx = Lm/m²
Resistance	Ohm	Ω	Ω = V/A
Conductance	Siemens	S	S = A/V

Some other Units and Conversions

	MKS system	CGS system	SI
Magnetic flux	1 Wb	10⁸ maxwells (M)	1 Wb
Magnetic flux density	1 Wb/m²	10⁴ gauss (G)	1 T
Magnetising force	1 At/m	4 × 10⁻³ oersted (oe)	1A/m
Magnetomotive force	1 At	0·4 π gilberts	1A
Energy	1 J	10⁷ ergs	1J
Conductance	1 mho (℧)	1 mho (℧)	1 S
Compliance	1 m/N	cm/dyne × 10⁻³	1 m/N
Compliance unit		1 cu = 10⁻⁶ cm/dyne	

APPENDIX

SYMBOLS USED IN CIRCUIT DIAGRAMS

On the following two pages is given a representative selection of symbols used in circuit diagrams, including all those most commonly found. It should, however, be pointed out that numerous minor stylistic differences from these symbols will be found in practice due to the slightly different conventions adopted by the many firms and organisations which publish circuits. Further differences will be found when examining circuits published abroad, although there is usually sufficient resemblance for their meanings to be clear. Symbols for more elaborate devices, such as microwave tubes, have not been included since these are generally identified as such.

COMPONENT SYMBOLS

Aerial	Dipole aerial	Earth	Chassis (not earth)	Piezo electric crystal	Cell	Battery	Metal rectifier or semiconductor diode

Inductance	H.F. transformer	Dust-cored inductance	Dust-cored transformer	Iron-cored inductance	Iron-cored transformer	Variable inductance	Preset inductance	Tapped inductance

Fixed resistance	Tapped resistance	Variable resistance	Potential divider (preset)	Ganged controls	N.T.C. P.T.C. Thermistors	Light-dependent resistance	Voltage-dependent resistance

Fixed capacitor	Feed-through (chassis) capacitor	Electrolytic capacitor	Fixed capacitors with common negative plate	Variable capacitor	Preset capacitor	Temperature-dependent capacitor	Voltage-dependent capacitor

On-off switch	Double-pole on-off switch	Double-pole double-throw switch	Multi-pole rotary switch	Push-button switch unit	Multi-way rotary switch	Changeover switch	Press-button switch (bell-push)

Electric bell	Photo-electric cell	Spark gap	Indicator lamp	Morse key	Relay	Vibrator	Jack-socket	Headphones

Wires connected	Wires crossing not connected	Screened lead	Twisted flex	Shorting link	Fuse	D.C. Direct current	A.C. Alternating current

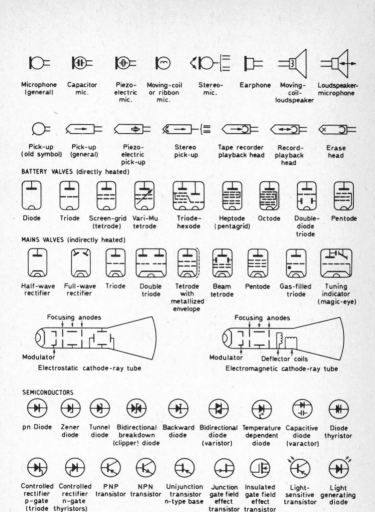

Microphone (general) | Capacitor mic. | Piezo-electric mic. | Moving-coil or ribbon mic. | Stereo-mic. | Earphone | Moving-coil-loudspeaker | Loudspeaker-microphone

Pick-up (old symbol) | Pick-up (general) | Piezo-electric pick-up | Stereo pick-up | Tape recorder playback head | Record-playback head | Erase head

BATTERY VALVES (directly heated)

Diode | Triode | Screen-grid (tetrode) | Vari-Mu tetrode | Triode-hexode | Heptode (pentagrid) | Octode | Double-diode triode | Pentode

MAINS VALVES (indirectly heated)

Half-wave rectifier | Full-wave rectifier | Triode | Double triode | Tetrode with metallized envelope | Beam tetrode | Pentode | Gas-filled triode | Tuning indicator (magic-eye)

Focusing anodes

Modulator

Electrostatic cathode-ray tube

Focusing anodes

Modulator | Deflector coils

Electromagnetic cathode-ray tube

SEMICONDUCTORS

pn Diode | Zener diode | Tunnel diode | Bidirectional breakdown (clipper) diode | Backward diode | Bidirectional diode (varistor) | Temperature dependent diode | Capacitive diode (varactor) | Diode thyristor

Controlled rectifier p-gate (triode thyristors) | Controlled rectifier n-gate | PNP transistor | NPN transistor | Unijunction transistor n-type base | Junction gate field effect transistor n-type channel | Insulated gate field effect transistor n-type channel | Light-sensitive transistor | Light generating diode

111

INDEX

114